Smart Energy and Electric Power Systems

Smart Energy and Electric Power Systems

Current Trends and New Intelligent Perspectives

Edited by

SANJEEVIKUMAR PADMANABAN

Department of Electrical and Electronics Engineering,
KPR Institute of Engineering and Technology,
Tamil Nadu, India

JENS BO HOLM-NIELSEN

Head of the Energy Section-Esbjerg, Head of the Center
for Bioenergy and Green Engineering, Department of
Energy Technology, Aalborg University, Esbjerg, Denmark

KAYAL PADMANANDAM

Associate Professor, BVRIT HYDERABAD College of
Engineering for Women, Hyderabad, India

RAJESH KUMAR DHANARAJ

Professor, School of Computing Science and
Engineering, Galgotias University, Uttar Pradesh, India

BALAMURUGAN BALUSAMY

Associate Dean-Student Engagement, Shiv Nadar Institute
of Eminence, Delhi-National Capital Region (NCR), India

ELSEVIER

Elsevier
Radarweg 29, PO Box 211, 1000 AE Amsterdam, Netherlands
The Boulevard, Langford Lane, Kidlington, Oxford OX5 1GB, United Kingdom
50 Hampshire Street, 5th Floor, Cambridge, MA 02139, United States

Notices
Knowledge and best practice in this field are constantly changing. As new research and experience broaden our understanding, changes in research methods, professional practices, or medical treatment may become necessary.

Practitioners and researchers must always rely on their own experience and knowledge in evaluating and using any information, methods, compounds, or experiments described herein. In using such information or methods they should be mindful of their own safety and the safety of others, including parties for whom they have a professional responsibility.

To the fullest extent of the law, neither the Publisher nor the authors, contributors, or editors, assume any liability for any injury and/or damage to persons or property as a matter of products liability, negligence or otherwise, or from any use or operation of any methods, products, instructions, or ideas contained in the material herein.

ISBN: 978-0-323-91664-6

For Information on all Elsevier publications
visit our website at https://www.elsevier.com/books-and-journals

Publisher: Charlotte Cockle
Acquisitions Editor: Graham Nisbet
Editorial Project Manager: Aera F. Gariguez
Production Project Manager: Anitha Sivaraj
Cover Designer: Matthew Limbert

Typeset by MPS Limited, Chennai, India

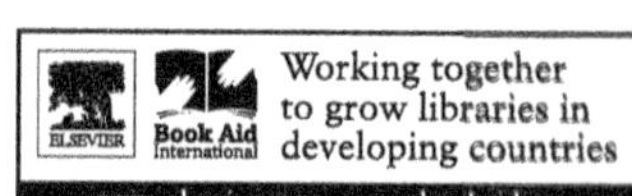

Contents

List of contributors

Aravind Prasad Baskaran
Department of Computer Science and Engineering, Saranathan College of Engineering, Trichy, Tamil Nadu, India

Arun Prasad Baskaran
Department of Computer Science and Engineering, Saranathan College of Engineering, Trichy, Tamil Nadu, India

Arturo Conde
Faculty of Mechanical and Electrical Engineering, Autonomous University of Nuevo Leon (UANL), San Nicolás de los Garza, Mexico

D. Renuka Devi
Department of Computer Science, Patrician College of Arts & Science, Chennai, Tamil Nadu, India

S. Rama Devi
BVRIT HYDERABAD College of Engineering for Women, Hyderabad, Telangana, India

Jorge Chan González
Faculty of Engineering, Autonomous University of Campeche (UAC), Campeche, Mexico

S.B. Gopal
Department of Electronics and Communication Engineering, Kongu Engineering College, Perundurai, Tamil Nadu, India

Ramya Gunasekaran
Department of Electrical and Electronics Engineering, Gnanamani College of Technology, Namakkal, Tamil Nadu, India

Iswarya Gururajan
Department of Computer Science and Engineering, Saranathan College of Engineering, Trichy, Tamil Nadu, India

V. Jayalakshmi
School of Computing Sciences, Vels Institute of Science, Technology and Advanced Studies (VISTAS), Chennai, Tamil Nadu, India

Manjushree Kumari Jayaraman
Department of Electrical and Electronics Engineering, Gnanamani College of Technology, Namakkal, Tamil Nadu, India

P.S. Latha Kalyampudi
BVRIT HYDERABAD College of Engineering for Women, Hyderabad, Telangana, India

D. Karthika
Department of Computer Science, Patrician College of Arts & Science, Chennai, Tamil Nadu, India

Rashmita Khilar
Saveetha School of Engineering, Saveetha Institute of Medical and Technical Sciences, Saveetha University, Chennai, Tamil Nadu, India

R. Senthil Kumar
Department of Electrical and Electronics Engineering, Sri Krishna College of Technology, Coimbatore, Tamil Nadu, India

D. Nanthiya
Department of Computer Technology-UG, Kongu Engineering College, Perundurai, Tamil Nadu, India

Lakshmi Kanthan Narayanan
Department of Computer Science and Engineering, Quannta School of Software Engineering, Villupuram, Tamil Nadu, India

Kayal Padmanandam
BVRIT HYDERABAD College of Engineering for Women, Hyderabad, Telangana, India

P. Pandiyan
Department of Electrical and Electronics Engineering, KPR Institute of Engineering and Technology, Coimbatore, Tamil Nadu, India

V. Pavithra
SRM Institute of Science and Technology, Chennai, Tamil Nadu, India

C. Poongodi
Department of Computer Science and Engineering, Vivekanandha College of Engineering for Women, Thiruchengode, Tamil Nadu, India

Hema Ramachandran
Department of Computer Science and Engineering, Saranathan College of Engineering, Trichy, Tamil Nadu, India

Nagarajan Ramalingam
Department of Electrical and Electronics Engineering, Gnanamani College of Technology, Namakkal, Tamil Nadu, India

R. Rengaraj Alias Muralidharan
Department of Information Technology, Saranathan College of Engineering, Trichy, Tamil Nadu, India

S.A. Sahaaya Arul Mary
Department of Computer Science and Engineering, Saranathan College of Engineering, Trichy, Tamil Nadu, India

N. Sai Charitha
BVRIT HYDERABAD College of Engineering for Women, Hyderabad, Telangana, India

Ranjani Sampathkumar
Department of Computer Science and Engineering, Saranathan College of Engineering, Trichy, Tamil Nadu, India

S. Saravanan
Department of Electrical and Electronics Engineering, Sri Krishna College of Technology, Coimbatore, Tamil Nadu, India

Meng Yen Shih
Faculty of Engineering, Autonomous University of Campeche (UAC), Campeche, Mexico

Venkatasubramanian Srinivasan
Department of Computer Science and Engineering, Saranathan College of Engineering, Trichy, Tamil Nadu, India

Priyanga Subbiah
Department of Networking and Communication, SRMIST, Chennai, Tamil Nadu, India

Kannadhasan Suriyan
Department of Electronics and Communication Engineering, Study World College of Engineering, Coimbatore, Tamil Nadu, India

Subetha Thangaraj
BVRIT HYDERABAD College of Engineering for Women, Hyderabad, Telangana, India

Ramji Tiwari
Department of Electrical and Electronics Engineering, Sri Krishna College of Engineering and Technology, Coimbatore, Tamil Nadu, India

Punitha Victor
Department of Computer Science and Engineering, Saranathan College of Engineering, Trichy, Tamil Nadu, India

Rajeswari Viswanathan
BVRIT HYDERABAD College of Engineering for Women, Hyderabad, Telangana, India

Francisco Lezama Zárraga
Faculty of Engineering, Autonomous University of Campeche (UAC), Campeche, Mexico

Preface

The economic growth of a country is mainly dependent on its primary infrastructures. One such primary infrastructure is electricity. Both industrial and service sector revolution have embarked the surge in demand for electricity. Developing countries have started investing to make electricity infrastructure smarter and began generating electricity from alternate sources such as wind, solar, biomass, etc. Despite the production of adequate electricity, the country still faces a huge demand for electricity, which leads to blackout.

Power outages are unpredictable and various factors contribute to it. Some of them include failure of components in local grid, network capacity overload, illicit connections to local grid, exposure to bad weather conditions, deliberate disconnection, etc. Load shedding can result in loss if power cuts are not planned well according to low demands. This cumbersome task of analyzing the demands for electricity unravels the technique of using artificial intelligence (AI) to smartly handle the demand of electricity without greatly affecting the economy of a nation.

Machine learning (ML) and AI techniques are extensively used in all the domains for the meticulous prediction of events by learning from the given training data. ML and AI techniques are extensively studied to improve accuracy in decision-making process. AI techniques can now be applied to power systems to predict the demand for electricity in various locations from the consumption history. This enables the power systems to be smarter in determining the load shedding schedule without impacting the economy. This can further predict the power outage events that are likely to occur from bad weather conditions, deteriorating grid components.

This book aims to bring the convenience of AI in power systems. This will address the issues associated with power systems and gives insight into the techniques of load shedding. AI techniques to predict the demand pattern are explored in detail. This book will enable researchers to focus on their ideas in improving the prediction capability of AI algorithms to make careful decisions in load shedding. Demand crisis in electricity cannot be eliminated, but it can be minimized with little impact by making intelligent power systems to handle the demands. This book also focuses on techniques and approaches leading to smart power systems.

CHAPTER 1

Smart power systems: an eyeview

Kayal Padmanandam[1], Subetha Thangaraj[1] and Rashmita Khilar[2]
[1]BVRIT HYDERABAD College of Engineering for Women, Hyderabad, Telangana, India
[2]Saveetha School of Engineering, Saveetha Institute of Medical and Technical Sciences, Saveetha University, Chennai, Tamil Nadu, India

1.1 Introduction to artificial intelligence

Many factors have contributed to the progress of AI and its sub domains. The first and foremost reason is the exponential growth of computing capacity. It has become available in large scale for training bigger and multifaceted complex models. This is possible and feasible due to the extra ordinary innovations made with silicon alike graphics processing units and tensor processing units, and many more on the way. This capacity is amazed with hyperscale clusters and made accessible to the user effortlessly through cloud architecture.

Despite the different artificial intelligence (AI) definitions alike, intelligent computer and machine technology, intelligent automated systems, intelligent machine that replaces human work, etc., it is built to help humankind by emulating cognitive abilities of human [1] to continue complex process hassle-free, when it is harnessed properly.

AI has invaded almost all the service oriented and business-oriented sectors in the universe, and so the power sector. Access to clean, cheap, and reliable energy is a fundamental right for human kind, which has direct impact on health, education, societal security, and livelihood. The main goal of sustainable development is to have universal access to such smart energy. Worldwide, researchers are involved in the automation of routines in the energy sector to prevent emergent markets from the nonexistence of satisfactory power generation, deprived transmission and distribution infrastructure, cost-effectiveness etc., Besides, the divergence, regionalization of energy production, changing demand patterns, crop intricate challenges for power generation, transmission, distribution, and consumption nationwide [2].

Smart Energy and Electric Power Systems
DOI: https://doi.org/10.1016/B978-0-323-91664-6.00006-1

AI has the capability to accelerate the use of energy sources in power grid in an efficient way. It can drastically reduce and monitor any kind of faults in the planning or procedure of large power system. It is the best approach for clean, cheap and reliable energy, which is a global need in today's scenario.

1.2 Necessity of artificial intelligence in power systems

Besides the swift growth in industrial development due to the advancement in power system expansion, there is also an exponentially increasing growing need for dynamic, stable, and reliable power system. Conventional power system techniques cannot handle this mass requirement due to the lack of modern infrastructure. Monitoring remote devices, heavy data acquisition and analysis are becoming complex and time-consuming process [3]. As the proverb says "Necessity is the mother of Invention," AI has become a necessity to solve the complexity in large power system.

Modern AI techniques like artificial neural networks (ANNs), expert system techniques (XPS), fuzzy logic (FL), genetic algorithm (GA) are utilized extensively in larger power system to manage complex task alike load dispatch, load forecasting, load optimization, transmission flow, generator monitoring, fault diagnosis, voltage control, electricity market analysis, security measures and very many operations and controls required for smart grids. These operations are handled intelligently through AI and can drastically reduce the human effort and risks involved in the power system (Fig. 1.1).

1.3 Modern artificial intelligence techniques in power system

The four major AI techniques used in power system are ANNs, XPS, FL, GA. These approaches can handle Intricate, versatile and huge information used for calculation, analysis and learning very easily. Despite vast data handling capabilities, they can also effortlessly handle higher computational system.

1.3.1 Artificial neural networks

ANN is a recent and successful technology applied in most of the complex and high priority systems. ANN is a human developed computing

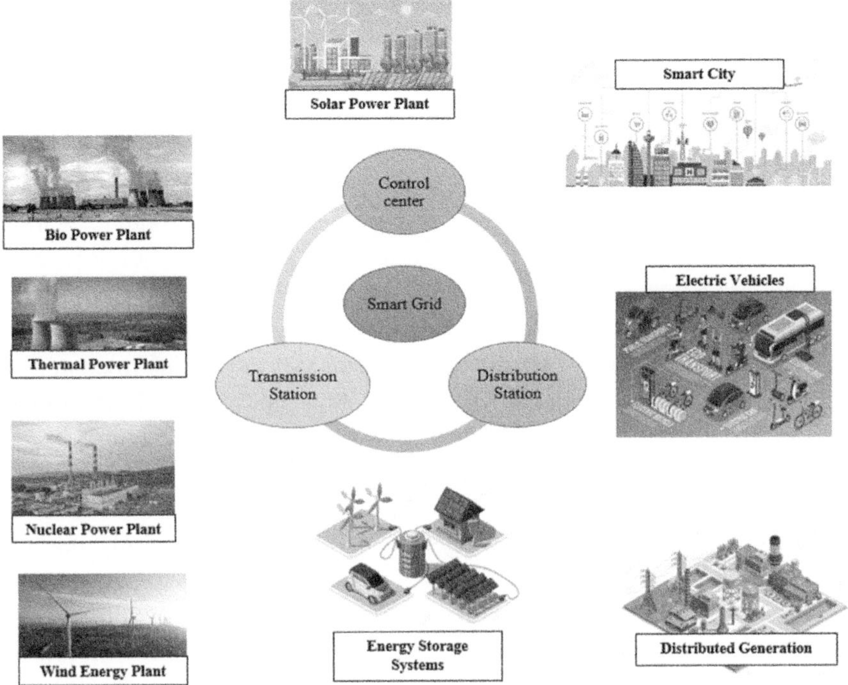

Figure 1.1 Necessity of AI in power systems. *AI*, Artificial intelligence.

system that emulates biological neural networks. It is a collection of interconnected nodes called neurons that simulates the electrical activity of a biological brain. These networked neurons convert set of inputs to set of outputs. Each neuron in the network, act as a processor, computes linear and nonlinear operations from the input, and produces an output through activation function. The complex interconnected working principle of neurons has the capability to solve many real-world problems because of its high end parallelism, asynchronous processing, multifrontal execution, real-time malleability, robustness toward damages and missing data, learning ability and involuntary generality, cutdown of additional software, fixed configuration and primarily the well-developed mathematical foundation [4,5].

ANNs are categorized based on the number of layers, topology, and the pattern of connectivity used. Their layers are input layer which represents the input vector, hidden layer which represents intermediary nodes, and output layer which represents output. Basically neural networks are classified into perceptron, feed-forward neural network, multilayer perceptron, convolutional neural network (CNN), recurrent neural network (RNN), radial basis functional neural network, long short-term memory (LSTM), sequence to sequence models, modular neural network, etc. Among these, the most important and most implemented networks are CNN, RNN, and LSTM. CNN are commonly used in image processing, computer vision and machine translation applications. They comprehend imageries in parts and compute the process manifolds to achieve the complete image processing output. Various applications in power system like event classification and localization [6], power network icing image [7] are efficiently monitored using CNN algorithms. RNN use sequential or temporal data and helps in prediction of a system. They need training data to learn and use "memory" to take information from preceding inputs and influence the current input and output. They can be used to predict the load forecasting [8], instability prediction etc., in smart grid applications. LSTM network is a cutting-edge RNN and a sequential network that consents information to persist. It is a highly celebrated for handling the vanishing gradient problem faced by RNN. LSTMs are explicitly designed with three gates forget, input and output to forget irrelevant information and carry over only relevant information, hence evade long-term dependency problems. They can be well utilized in smart grid for energy management [8], stability prediction [9], demand forecast, network delay estimation [10] etc., efficiently.

1.3.2 Expert system techniques

An expert system [11] in AI also known as knowledge-based system that mimics the concluding capability of a knowledgeable human. The expert system replicates these knowledges and utilize it to solve analogous problems without human expert participation. Various extents of applications in power systems where a massive volume of data processing has to happen rapidly matching the ability of expert systems. As they are computer-based programs, the procedure of scripting codes is simpler than calculations and estimations used in generation, transmission and distribution of power system. It is dynamic and paves way for virtual estimations and even modifications after designing the programs (Fig. 1.2).

1.3.3 Fuzzy logic (FL)

FL approaches are different from modern computing. It depends on degrees of truth rather than the binary Boolean True (1) or False (0). It has a range of value where 1 and 0 acts as extremities with various

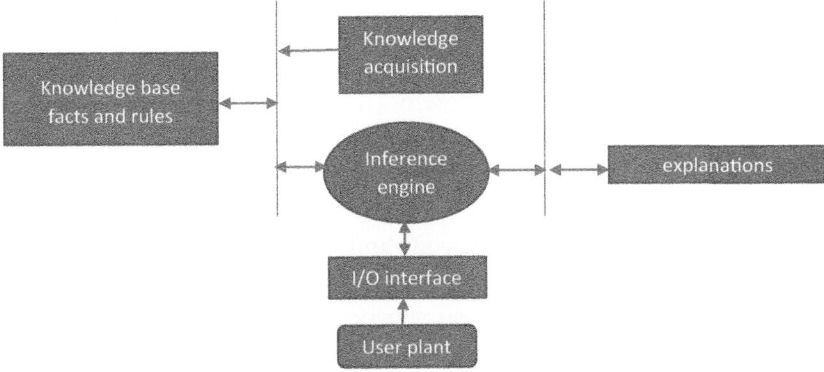

Figure 1.2 Structure of expert systems.

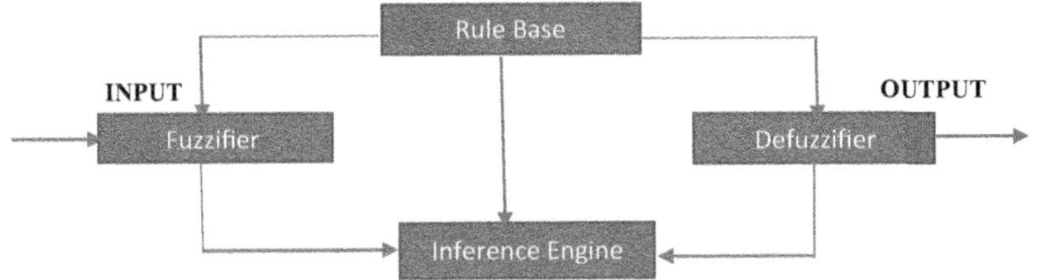

Figure 1.3 Fuzzy logic architecture.

intermediate degrees of truth. This feature helps in many engineering applications where the data is uncertain and imprecise. Many real time systems like temperature control, large scale power system control, system that demands reliability and safety control highly relies on FL engineering. Even, the rising complexity in power systems necessitate the application of FL in many power system glitches (Fig. 1.3).

1.3.4 Genetic algorithm

A GA [12] is an exploring experiential optimization technique that imitates the procedure of selecting the fittest individuals from a pool of individuals thereby giving a higher chance to yield more "fitter" individuals. It is efficiently used for solving optimization problems in machine learning that would take a long time to solve. It is a highly used in varied fields of power system applications such as Wind turbine positioning, reactive power optimization, maintenance scheduling, capacitor placement, network feeder routing, loss minimization, load management, load frequency control, load flow and many more.

1.4 Artificial intelligence techniques in smart grids

The power systems have evolved from the typical energy systems to a powerful posterity smart grid system [13]. The main principle of smart power grid works on substituting the labor-intensive maneuvers with artificial intelligence to attain the convenience of high efficacy, fidelity, and low outlay. The utilization of AI in smart grids likely cut down the energy waste, lessen the energy cost, and expedite the utilization of sustainable energy sources in worldwide power grids. The power systems planning, activity, and control can also be improved. The inclusion of AI technologies in power systems contributes cheap energy that is indispensable to the progression. The most prominent places where AI is applied are hydroelectric power, solar power plants, smart houses, smart industries, smart cities, electric vehicles, energy storage systems, distributed generation, power plants, etc. This section describes the most widely used AI applications, including load forecasting, faults detection, power grid stability assessment, and smart grid security. The Smart Grid AI applications are shown in Fig. 1.4.

1.4.1 Load forecasting

Load forecasting [14] is a substantial actor for the generation of electric energy, marketplaces, automatic transmission as well as distribution. This

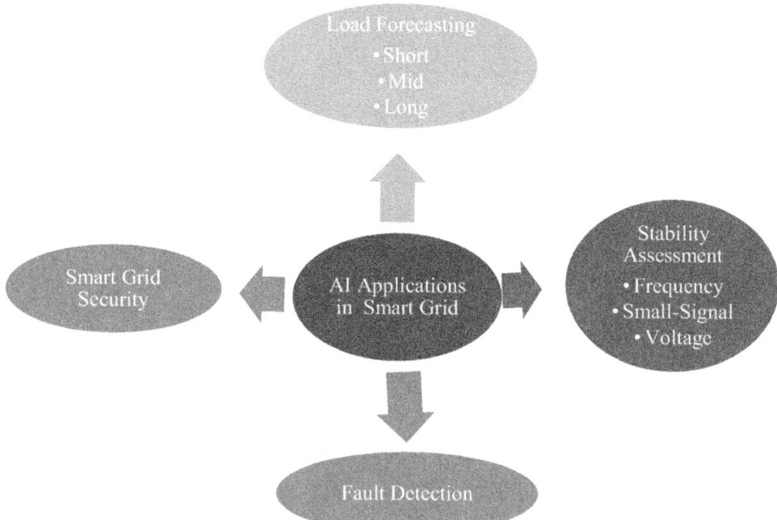

Figure 1.4 AI applications in smart grid. *AI*, Artificial intelligence.

plays a vital role in electricity generation planning and operation. The load forecasts [15] assist in determining the approaches to enhance the operating mechanisms in a driven period, thereby ensuring the appeal during the contrary effects. Load forecasting is mainly needed for transmission planning and distribution, power grid planning and operation, development of manpower, and electric sales. Moreover, the pertinent forecasts can substantially increase the revenue of the company by reducing the operational and maintenance cost. It also improves the consistency of power supply and delivery thereby leading to suitable verdicts for future predictions. Thus, the research area in load forecasting is growing due to the advancements and inclusion of AI in smart power systems. The AI in smart power systems reduces the errors in prediction and it is an indispensable part of both power systems and energy management systems. The major challenges of load forecasting are seasonal effects, unpredictable weather conditions, and captured load signals. Hence, Efficient techniques are needed to achieve excellent management, operation, and planning of electric power systems.

In general, the generation and demand of energy are predicted using a prediction model, and they can be categorized into two categories [16]. With the assimilation of smart grid systems through regular power systems and the inflating addition of renewable energy sources, accurate electric load forecasting becomes more intricate. The first one is based on the application of building a statistical method, and the second one is based on Machine Learning, and some systems also utilize the combination of statistical models and machine learning algorithms. The machine learning algorithms are grouped into three techniques such as trend analysis, correlation and combination of both to estimate the energy demand. The first technique trend analysis utilize the previous energy requirement data and use those data to predict the future energy demand. This uses regression algorithm to predict the future points with the help of historical data. The trend analysis works well in real-time scenarios. The next technique namely correlation techniques [17] associate various factors such as economic and demographic data to extract the association between increase patterns of load and various factors influencing the energy demand. The main issue affecting the performance of correlation techniques is the prediction of various factors taken into consideration because it is much difficult than the prediction of load forecast [18].

Load forecasting necessitate proper strategy and practical applications require demanding "forecasting intervals," also mentioned as "lead time."

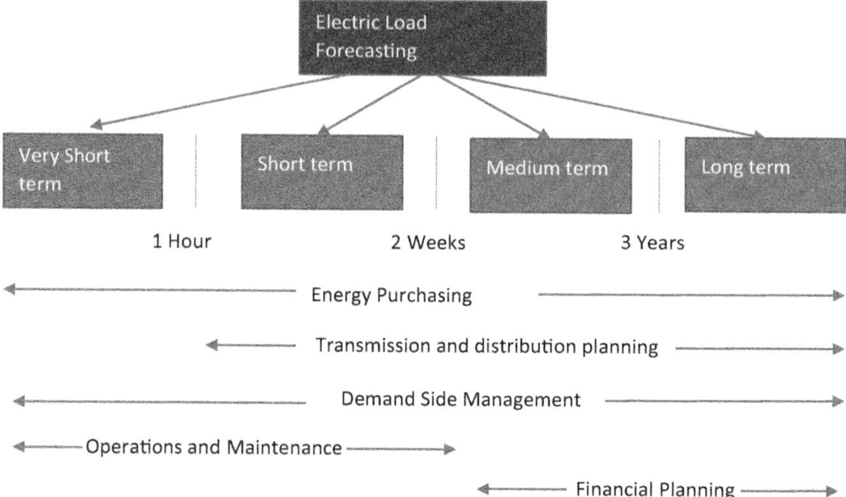

Figure 1.5 Electric load forecasting applications and classifications.

Load forecasting strategies [19] are clustered into four categories such as Very-short-term forecasting, short, medium and long-term models based on the periods. Very-short-term forecasting spans only for few minutes. Short-term forecasting spans between 1 h to a month and this is mainly used in scheduling day-to-day operations in utility industry. Medium Term Load Forecasting spans between 1 week and 1 year. Long-term forecasting spans from few years to few decades and it is applicable for predicting the new generations construction, strategic planning and switch based on the energy supply and distribution. Fig. 1.5 shows detailed illustrations about load forecasting classifications.

1.4.2 Power grid stability analysis

The most demanding need for a smart grid is to have a secure and stable operation. But the advanced features have created stability features considerably extra multifaceted than that of traditional power grids. The traditional stability analysis and control techniques are not effective in certain circumstances. To circumvent these situations, AI can confront these modifications and supply assuring tools for ensuring the safety and stability requirement [20]. Many researchers started working toward the interdisciplinary field on AI applications in stability control and analysis. The major applications in this area are security and stability assessment, stability control, and fault diagnosis. The AI research in this field has been gaining significant attention due to improved

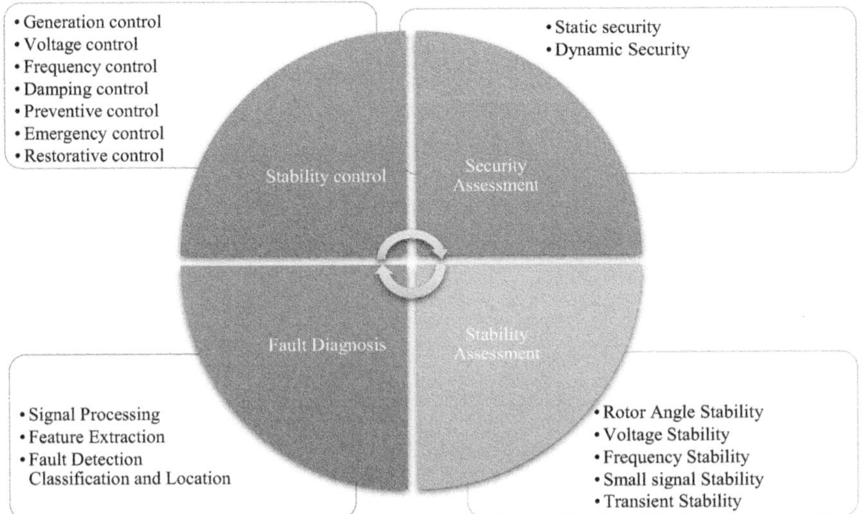

Figure 1.6 AI applications in stability analysis. *AI*, Artificial intelligence.

performance criteria such as accuracy, efficiency, and speed. The overall summarized applications are shown in the Fig. 1.6 [17].

Stability analysis intends to examine whether the power grid can uphold the stability beneath the disruption and present the instability type. Suitable measures can be taken when we determine the instability type. Nevertheless, with the development of interrelated power grids, an amalgamation of renewable energy, and the commissioning of High-Voltage Direct Current (HVDC), it becomes burdensome for conventional techniques to handle the intricate features of the scheme.

At present, the transient energy function (TEF) and time-domain simulation (TDS) techniques are frequently used. Despite greater preciseness, TDS is time-consuming to encounter the necessities of online stability analysis. In addition, intricate models with exact parameters need to be constructed to resemble real power grids. The energy function is challenging for enterprise power grids in the TEF technique, and the widely available TEF models neglect the complex transitory process. This leads to affect the preciseness of stability analysis and the overall.

1.4.3 Stability assessment

With the continuous load demands, there is an inflation of utilizing renewable energy, various equipment's, smart grid developments thereby making stability a deliberate issue. Thus, detecting early instability

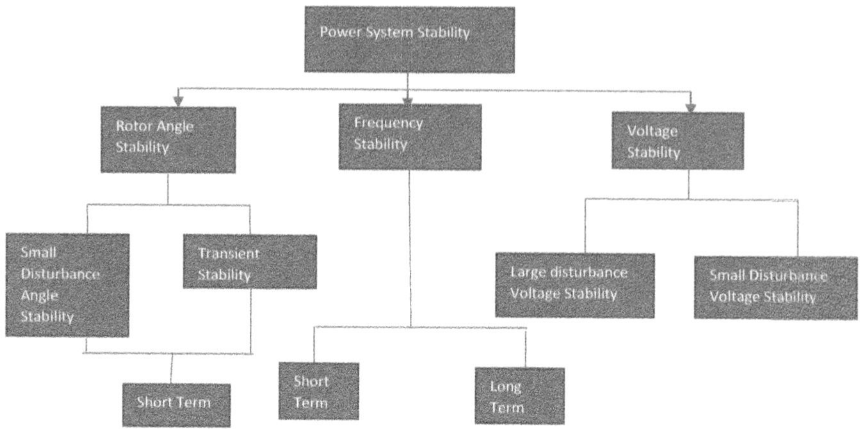

Figure 1.7 Power system stability classification.

performs a vital role for maintain safe and balanced activity in smart grids. Based on IEEE/CIGRE Joint Task Force, the stability problem can be categorized into rotor angle stability, frequency stability, and rotor angle stability. To improve the speed, performance, and accommodate, the researchers started to adapt AI in stability assessment. This section discusses about the three types of stability assessment and the overall classification is shown in Fig. 1.7.

1.4.4 Rotor angle stability

Rotor angle stability assigns capability of contemporaneous machines of an interrelated power system stays in contemporaneous even after it gets disturbed. When the generators are running alongside and transfer the power relies on the rotor angle. The rotor magnetic field and stator magnetic field spins in an unchanging speed for normal procedure. But, the angular separation count on the generator power output. The rotor angle will be fixed in the fresh location analogous to the revolving magnetic field with an escalation of turbine speed. However contraction in the torque recede the rotor angle proportionate to the stator field. In general, the stability remains same for equilibrium condition and if the equilibrium is agitated there will be either acceleration or deceleration in the machine rotors. Two generators are running parallelly and if one generator temporarily progresses in a rapid manner in comparative to other then the machine's rotor angle will progress in the direction of slow machines.

Hence, the faster generator distribution of load increases and vice versa. This will also lead to the power transfer decline.

1.4.5 Voltage stability

Voltage stability also known as load stability is used to determine the capability to manage adequate voltages to all the bus under regular state and even after being exposed with commotion. The voltage of the system will remain stable during normal operating conditions and if voltage turns into unstable condition there is an intractable deterioration in voltage. Voltage instability may lead to some voltage collapse. Voltage collapse is also known as voltage instability. This can be grouped into two classes. They are large and small disturbance voltage stability. Prior voltage stability belongs to the extensive disruption like load loss, and failure of generation whereas later disturbance voltage stability have less disturbance.

1.4.6 Frequency stability

Owing to the high demands in loads, Frequency stability plays a crucial role in organizing and planning. The balance in load-generation has to be done in each power system to prevent the damage. Frequency Stability is defined as the capability of a power system to uphold stable frequency ensuing drastic explosion betwixt load and generation. This also based on the strength to bring back evenness between load and generation with manageable load loss. The main issue with the instability is the continual change of frequency which leads to tripping. Though classification of stability is grouped into frequency, rotor and voltage, it is not mandatory to be independent. For example, if there occurs a voltage drop in bus, it will also affects rotor and frequency components. In the same way, higher frequency aberration drives to voltage distortions. Hence, all the three has to be protected to maintain stability.

1.4.7 Small signal stability

Small signal stability is defined as the system's capacity to maintain synchrony even in small commotion. Small-disturbance angle stability analysis is the inspection of the rotor angle stability where the commotion is limited bounteous so that nonlinear power system can be proximate by a linear system. A few examples of small commotions are tiny load changes such as tiny load interchange of on or off, tripping, etc. oscillatory and non-oscillatory instability are two ways of categorization. The generator's rotor angle tends to grow owe to small disturbances for non-oscillatory instability and in the other case the rotor angle oscillates with increasing magnitude.

1.4.8 Transient stability

Transient Stability can be defined as the system's capacity to maintain synchrony in large commotion. A few examples of big commotions are big load changes such as big load interchange of on or off, tripping, etc. In general, the difference between small and large disturbance occurs in between the time gap of 0.1−10 s.

1.4.9 Stability control

As discussed in the earlier sections, power system stability endures to stay in symmetry in both typical and agitation state. The entire world is struggling to balance the power stability mainly due to two whys and where foes.
1. Reserving the transmission for various conditions thereby functioning in little security margins.
2. Reforming the power systems in such a way that is not concurrent further.

As a result, the power system stability decreases leading to shutdowns. To control these situations, it is essential to have power system stability control to maintain the system efficiently. Various researchers are working toward utilizing phasor measurement unit to control the power system. Some researchers are working toward utilizing the new advancements in technologies to make an expert system for control. Fig. 1.8 shows the

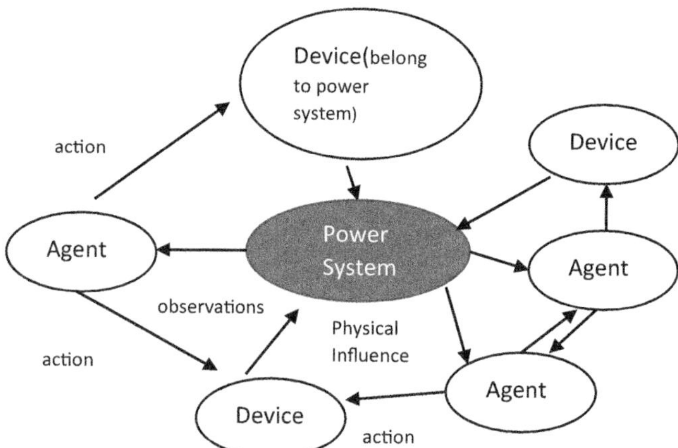

Figure 1.8 Power system stability control.

power system stability control where the perception are accomplished and disseminated to other handler to control applicably the liable devices.

1.4.10 Security assessment

Security assessment [21] is the capability of a power system to persevere the typical operations despite unintended expenses to the possibilities. The deterioration of security can produce a minimum loss in device destruction to the maximum syncope. The main aim of security assessment is to calculate the susceptibility rate of the device to become syncope. Despite various research, the traditional techniques fail because they are expensive in time and space. Modern-grids are handled in real-time by adopting Machine learning and IoT Techniques.

1.4.11 Fault diagnosis

Another major requirement for a power system is fault diagnosis [22] which can be achieved in real-time by combining with AI. In general, the failures of machines will lead to resource wastage, amount, duration and lives. With the advancement of AI fused with the sensors, the failures of the devices can be identified in the early stages thereby utilizing the resources without any wastage. Geothermal energy can be used as a main source of power to aid in power. Several companies are working to incorporate IoT and AI to revamp the renewable energy sources. Nowadays researchers started utilizing predictive diagnostics to identify and forecast the alleviations in the earlier stage to refrain shut down. The precautionary steps such as synthetic spray are performed to optimize the performance.

1.5 Challenges in smartgrid

Grid infrastructure is still emerging in developing among nations like India. The present grid network is insufficient to meet the anticipated demands for clean energy and distributed generation, which may pose a numerous issue in terms of construction, operation, design and maintenance [23]. Besides smart grid attentiveness, there is also a need to solve faults with the current grid infrastructure [24]. Several electrical sections of India are not even linked to the national grid to optimally evacuate enormous wind farms or solar parks, which will demand the expansion of complete infrastructure of smart grid [25,26].

The smart grid will meet these requirements effectively [27]. Advances in information and communication technologies help to support the areas related to smart grid idea. Sensors and control technology also helps in the conversion of an already existing grid to a smart grid technology.

Various issues exist in this area, among those some of the issues stated in the field of smart grid includes the following: the freedom system, intelligence, communications, integrating the intermittent generations, moving offshore, capturing DG advantages and storage, and preparing plugin hybrid cars. Let us identify each one independently [28]. Grid capacity should be guaranteed to interconnect energy resources, mostly renewable ones. The electric power grid has been in operation for over a century and is often regarded as the world's largest and most complicated linked physical system. It is referred described as an ecosystem because of its breadth, complexity, and intrinsically related to human evolution and activity [27]. The Smart System is a technological difficulty that extends much beyond just putting an IT architecture on top of an electrical grid [29]. Every device linked to a Smart Grid serves as both an electrotechnical device and an intelligent node.

1.5.1 Smart grid security challenges and objectives

The design and infrastructure of SGs confront a plethora of security risks and problems, including theft, cyber-attacks, terrorism, natural catastrophes, and so on. Power system blackouts (small and big outages), SG IT infrastructure failures, inaccurate visualization of the actual system's state, cascade failures, damaged consumer products, energy market pandemonium, jeopardized human safety, and other effects are likely if SG fails owing to any of the threats [26,30].

Meter Infrastructure Threats and vulnerabilities smart meters in smart grid infrastructure house the most useful data (e.g., meter readings) for enhancing power grid performance and transforming the lives of energy users. Numerous smart grid applications and services, such as automatic meter reading, invoicing, dynamic pricing, and warning of approaching blackouts and energy thefts, need readings, which may be very convenient for both utilities and energy users.

1.5.2 High-level security requirements

Electric Power Research Institute (EPRI) perhaps the greatest test looking in the shrewd framework sending is identified with digital protection.

Digital protection is a basic issue due to the in wrinkling capability of cyberattacks [31]. Further more episodes against this area as it turns out to be more interconnected. Network protection should address not just intentional assaults, as modern secret activities, or fear mongers, however coincidental tradeoff of the data foundation because of human mistakes, gear disappointments, and catastrophic events.

1.5.3 Quality of service-related requirements

The association between the power provider to the clients is a principle issue of shrewd framework, and it is important to ensure quality of service for the correspondence and systems administration in every one of the phases of the Smart Grid, going from power age, transmission, appropriation, to the client applications. guarantee that the observing information, crisis reaction and control order can be conveyed inside required timeframe, yet it ought to be impacted by the quantity of PDA client information traffic [32,33].

1.5.4 Transmission system

SCADA is the core of transmission framework at load dispatch focuses, which is presently applied to bigger and widespread organizations (WANs) because of headway in data innovation. Despite heap dispatch focuses have autonomous control, they are likewise associated with different focuses through a typical correspondence framework. Today many significant transmission organizations have coordinated web into the correspondence network for better effectiveness and unwavering quality [28]. However, this puts the whole framework at significant danger as an enemy figures out how to penetrate any of the SCADA organizations, then, at that point, it can make devastation the whole framework activity [34].

1.5.5 Distribution system

A regular meter can be adjusted by switching the interior use counter or can be controlled to control the estimation of electric stream. Canny electronic devices (IED) like savvy meters can be controlled to convey different functionalities from distant area. This empowers a foe to remotely associate or disengage the gadgets or mess with information shipped off the framework administrator or slip into secret information of the buyers. Likewise, in case a foe figures out how to send bogus information bundles to infuse negative estimating in the framework, then, at that point, it will bring about power deficiencies at the designated region making loss of income the service organization [25]. Considering that there are a great many regular/shrewd meters associated with the framework, it is hard to get each hub expanding the weaknesses of the framework to complex occasions. Anderson and Fluorita have shown that an assailant could turn off large number of shrewd meters at the same time through a faroff area. Savvy meters likewise neglect to agree to the Open Web Application Security Project (OWASP) guidelines like infusion, validation, cross web-page prearranging (XSS), shaky direct article references [25].

1.5.6 Telemetry infrastructure

Telemetry frameworks are regularly dismissed during security arranging, testing, and assessment process. They associate with control frameworks and SCADA engineering of different parts in brilliant lattice: age frame-works, transmission frameworks, dispersion frameworks, and miniature matrices.

- A disappointed representative who has advantage to get to the frame-work parts may modify the calculations of programming or alter the settings of gadgets causing misleading activity. Additionally, corporate information can be taken from the data set for inward competition between the contending specialist organizations. Aggressor can utilize key lumberjack programming to get close enough to framework user-names and passwords. Such activities may not exclusively be hard to distinguish yet in addition to forestall.
- Some gadgets like PC's and USB flash drive that are utilized both within and farther the believed edge can be tainted with malware, and afterward attack the framework when utilized inside [28].
- Gadgets can likewise be negotiated prior to conveying them to the site by varying the code of programming or altering the control settings of equipment. Such assaults are inventory network assaults [28].

References

[1] S.J. Russell, P. Norvig, Artificial Intelligence: A Modern Approach, Prentice Hall, Upper Saddle River, NJ, 2009 [Google Scholar].
[2] Fresh Ideas about business in emerging markets, IFC, a member of the World Bank Group, Note 81, April 2020.
[3] https://www.electricalindia.in/artificial-intelligence-an-advanced-approach-in-power-systems/.
[4] S. Walczak, N. Cerpa, Artificial neural networks, Encyclopedia of Physical Science and Technology, third ed., 2003, pp. 631−645. ISBN 9780122274107.

[5] I. Kononenko, M. Kukar, Artificial neural networks, Machine Learning and Data Mining (2007) 275−320. ISBN 9781904275213.

[6] H. Ren, Z.J. Hou, B. Vyakaranam, H. Wang, P. Etingov, Power system event classification and localization using a convolutional neural network, Frontiers in Energy Research 8 (2020) 607826. Available from: https://doi.org/10.3389/fenrg.2020.607826,oct2020.

[7] J. Lu, Y. Ye, X. Xu, et al., Application research of convolution neural network in image classification of icing monitoring in power grid, Journal on Image and Video Processing 2019 (2019) 49. Available from: https://doi.org/10.1186/s13640-019-0439-2.

[8] J. Zheng, C. Xu, Z. Zhang, X. Li, Electric Load Forecasting in Smart Grid Using Long-Short-Term-Memory based Recurrent Neural Network.

[9] M. Alazab, S. Khan, S.S.R. Krishnan, Q. Pham, M.P.K. Reddy, T.R. Gadekallu, "A multidirectional LSTM model for predicting the stability of a smart grid,", IEEE Access 8 (2020) 85454−85463. Available from: https://doi.org/10.1109/ACCESS.2020.2991067.

[10] R. Feizimirkhani, H. van Nguyen, Y. Bésanger, Q. Tran, A.I. Bratcu, et al., Application of long short-term memory (LSTM) neural network for the estimation of communication network delay in smart grid applications, in: EEEIC 2021 − 21st IEEE International Conference on Environment and Electrical Engineering (EEEIC 2021), September 2021, Bari, Italy. ⟨hal-03357785⟩.

[11] Z. Zhang, G.S. Hope, O.P. Malik, Expert systems in electric power systems − a bibliographical survey, Power Engineering Review, IEEE 9 (1989) 33. Available from: https://doi.org/10.1109/MPER.1989.4310368.

[12] J. McCall, Genetic algorithms for modelling and optimisation, Journal of Computational and Applied Mathematics 184 (1) (2005) 205−222. Available from: https://doi.org/10.1016/j.cam.2004.07.034.

[13] C. Bharathi, D. Rekha, V. Vijayakumar, Genetic algorithm based demand side management for smart grid, Wireless Personal Communications 93 (2) (2017) 481−502. Available from: https://doi.org/10.1007/s11277-017-3959-z.

[14] A. Singh, M. Kumar Srivastava, An overview of artificial intelligence techniques for efficient load forecasting, ARPN Journal of Engineering and Applied Sciences 14 (9) (2019) 1800−1808.

[15] S. Shrivastava, K.T. Chaturvedi, A review of artificial intelligence techniques for short term electric load forecasting, International Journal of Advanced Research in Electrical, Electronics and Instrumentation Engineering 7 (5) (2018) 2241−2247.

[16] M.Q. Raza, A. Khosravi, A review on artificial intelligence based load demand forecasting techniques for smart grid and buildings, Renewable and Sustainable Energy Reviews 50 (2015) 1352−1372.

[17] B.K. Bose, Artificial intelligence techniques in smart grid and renewable energy systems—some example applications, Proceedings of the IEEE 105 (11) (2017) 2262−2273.

[18] E. Vivas, H. Allende-Cid, R. Salas, A systematic review of statistical and machine learning methods for electrical power forecasting with reported mape score, Entropy 22 (12) (2020) 1412.

[19] I.K. Nti, S. Asafo-Adjei, M. Agyemang, Predicting monthly electricity demand using soft-computing technique, International Research Journal of Engineering and Technology 06 (2019) 1967−1973.

[20] Eeeguide.com, Forecasting Methodology (2014). Available from: http://www.eeeguide.com/forecasting-methodology/. Accessed 4 Jan 2019.

[21] Z. Shi, W. Yao, Z. Li, L. Zeng, Y. Zhao, R. Zhang, et al., Artificial intelligence techniques for stability analysis and control in smart grids: methodologies, applications, challenges and future directions, Applied Energy 278 (2020) 115733.

[22] N.V. Tomin, V.G. Kurbatsky, D.N. Sidorov, A.V. Zhukov, Machine learning techniques for power system security assessment, IFAC-PapersOnLine 49 (27) (2016) 445−450.

[23] B. Makala, T. Bakovic, Artificial Intelligence in the Power Sector, 2020.

[24] A. Bari, J. Jiang, W. Saad, A. Jaekel, Challenges in the smart grid applications: an overview, International Journal of Distributed Sensor Networks (2014). Available from: https://doi.org/10.1155/2014/974682.

[25] C.P. Vineetha, C.A. Babu, Smart grid challenges, issues and solutions, 2014 International Conference on Intelligent Green Building and Smart Grid (IGBSG) (2014) 1−4. Available from: https://doi.org/10.1109/IGBSG.2014.6835208.

[26] A. Kerckhoffs, La cryptographie militairie, Journal of Sciences Militaires IX (1883) 5−38.

[27] A.H. Ahangar, H.A. Abyaneh, G.B. Gharepetian, Negative effects of cyber network (control, monitoring, and protection) on reliability of smart grids based on DG penetration, in: 2015 5th International Conference on Computer and Knowledge Engineering (ICCKE), 2015, pp. 54−60. Available from: https://doi.org/10.1109/ICCKE.2015.7365861.

[28] E.L. Quinn, Smart metering and privacy: Existing laws and competing policies, SSRN eLibrary (2009).

[29] J. Vijayan, Stuxnet renews power grid security concerns, Computerworld (2010).

[30] Y. Mo, T.H. Kim, K. Brancik, D. Dickinson, H. Lee, A. Perrig, et al., Cyber-physical security of a smart grid infrastructure, Proceedings of the IEEE 100 (1) (2012) 195−209.

[31] Y. Huang, M. Esmalifalak, H. Nguyen, R. Zheng, Z. Han, H. Li, et al., Bad data injection in smart grid: attack and defense mechanisms, IEEE Communications Magazine (2013) 27−33.

[32] E. Bou-Harb, C. Fachkha, M. Pourzandi, M. Debbabi, C. Assi, Communication security for smart grid distribution networks, IEEE Communications Magazine (2013) 42−49.

[33] P. Mohajerin Esfahani, M. Vrakopoulou, K. Margellos, J. Lygeros, G. Andersson, Cyber-attack in a two-area power system: impact identification using reachability, Proceedings of the American Control Conference (2010) 962−967.

[34] Electric Power Research Institute, Report to NIST on Smart Grid Interoperability Standards Roadmap, 2009.

CHAPTER 2

Smart energy and electric power system: current trends and new intelligent perspectives and introduction to AI and power system

V. Pavithra[1] and V. Jayalakshmi[2]
[1]SRM Institute of Science and Technology, Chennai, Tamil Nadu, India
[2]School of Computing Sciences, Vels Institute of Science, Technology and Advanced Studies (VISTAS), Chennai, Tamil Nadu, India

2.1 Introduction

Many countries are adopting artificial intelligence (AI) these days as it plays an essential part in numerous sophisticated technologies and also helps to integrate diverse sectors such as education, health care, water and environment, finance, and information technology. This chapter concentrates entirely on the many aspects of AI and power systems. Working with the modern technology presented several challenges for traditional ways. Throughout modern systems, AI has addressed every operational challenge. The power system's primary purpose is to supply high-quality electric power at a reasonable cost while ensuring reliability and consistency. Because the existing method in the computer program employs a numerical method, achieving a target for the power system is highly challenging, particularly under fault conditions [1]. The workload of the employees grows as the power system develops. AI techniques are used to address these issues, resulting in increased efficiency, dependability, and consistency. As a result, numerous researchers are working on AI applications for power systems. IN power system involves distribution, generating power, electric power utilization, etc. This traditional technique is challenging to adopt a new approach because of its complexity [2]. Smart grid in AI refers to a hybrid of current information systems and traditional power grids. The development of power systems is moving in the

Smart Energy and Electric Power Systems
DOI: https://doi.org/10.1016/B978-0-323-91664-6.00001-2

direction of the smart grid. It can address low energy efficiency, poor interface, and challenging security and stability analyses that plague traditional power systems. The structure and makeup of today's power grid are becoming increasingly complicated.

2.2 Power system

The power system is a vast network that is used for the generation of power, distribution of the power, and transmission system used for distribution. The emphasis is generated by various inputs such as coal, diesel, water, and solar energy. Power systems use multiple devices like circuits, motors, transformers to generate power. The power is generated from the power plant, and it is transformed to the transformer by a step-down approach. The emphasis is transferred to the substation and then to the industries or home system from the system.

2.3 Overviews of artificial intelligence

The world is changing rapidly, and we are being bombarded with new technologies daily. AI is one of the significant areas in the field of research. AI creates new technology all around the world by making intelligent machines. AI plays a crucial role in today's environment. It is currently used in several applications such as painting, self-driving automobiles, provincial theorems, etc. AI can have human-like abilities such as learning, problem-solving, and reasoning [3]. Even if you design a machine with a programmed algorithm, it is not required to pre-program it while using AI. AI is a cognitive approach that aims to further research. It is a replica of human behavior displayed by machines and computer systems. AI has been increasingly important in sectors. Such as information technology, brain research, medical data, and neurology in recent years. AI refers to intelligence processes performed by a computer or machine to complete a complex assignment quickly and efficiently. Traditional AI approaches include primary research, programming automation, natural language processing, problem-solving, and theorem-proof. These methods are very successful at completing tasks, but they have a few limitations [4]. In addition, traditional AI is not adept at handling complicated logical and dynamic thinking problems [5]. A new AI-based on neural network is developing to address these concerns, which, when paired with traditional AI, helps alleviate all of the issues traditional AI has. Expert systems, machine vision, fuzzy, artificial neural networks (ANNs), and natural language processing are examples of modern AI. The chapter with the power system application focuses on these AI approaches.

2.4 Applications of artificial intelligence

AI is now widely used in a variety of fields. AI can tackle complex problems quickly and efficiently, making life easier for humans.

2.4.1 Artificial intelligence in health care

People all over the world have been impacted by numerous diseases these days, and it is really essential to predict the conditions at a preliminary phase. A large amount of data are handled by healthcare organizations, and it can be challenging to predict diseases on time. With the assistance of AI, doctors can diagnose diseases more quickly and alert when a patient's conditions deteriorate, allowing medical assistance to reach the patient before hospitalization. The healthcare industry has shown to be a leader when it comes to AI. This is understandable given that those working in this field are presumed to save person's life as fast and efficiently. They don't see AI as a game; they need it, and it shows. Data management and analysis are two well-known AI applications in healthcare. Thanks to the Electronic Patient Record and machine learning (ML), caregivers can now predict disease outcomes based on the information regarding signs, trends, and style of living. This technique not only improves their own patients' treatment regimens, and ability to strengthen public health in general. Foremost, neural network is being integrated into an increasing number of devices, such as fitness trackers and insulin testing devices. Not to mention the numerous medical applications that provide consumers with health advice without the need for a doctor's intervention.

2.4.2 Artificial intelligence in robotics

AI plays a big role in the field of robotics. AI sense the real-time problem in its way and plan according to that. It gives learning capabilities, flexibility, in the previously rigid application. AI can be used for
- It helps to clean the office equipment.
- It helps to carry goods in warehouses, hospitals, offices.
- It helps in inventory management.

2.4.3 Artificial intelligence in gaming

In the gaming industry, AI is used to create human-like NPCs that interact with players. It is used to predict human behavior, which aids in the design of the game and testing. For example, it aids in establishing rules for games such as checkers, chess, and so on, and it provides the rules in a structured manner. Which aids in the improvement of learning abilities and problem-solving, achieving a certain level of performance.

2.4.4 Artificial intelligence in e-commerce

2.4.4.1 Personalized shopping

AI assists customers in personalized shopping based on their interests, preferences, and purchasing history. It establishes a recommendation engine with the customers and aids in the development of a relationship with them.

2.4.4.2 Fraud detection

AI solves significant problems in e-commerce applications such as credit card fraud and fake reviews. It aids in the reduction of the problem faced by e-commerce companies, as well as the removal of credit card fraud. Customers nowadays buy products based on the reviews left by other customers. Sometimes there are fake reviews on the site; AI solves this problem.

2.4.4.3 Artificial intelligence in recruitment

Because specific skills are required for particular jobs, industries such as IT suffer from understaffing. Recruiters do everything maximum to find the best and smartest IT specialists, but they have limitation due to limited resource and time; however, AI eradicates this need. To fit the right person with the right company, an intelligent computer can evaluate massive amounts of company information, culture standards, employing trends, and career opportunities. Furthermore, intelligent computer systems can assess recent successful job titles and use this data to create the overall job offer. Even though computers' writing abilities leave something to be desired, the final product still needs human eye.

2.5 Types of machine learning

ML is a subset of AI, which enables software applications by sensing trends and making intelligent decisions based on approved data samples. The majority of ML algorithms work with previous data as input data, and new output values can be predicted. It allows software programmers to predict outcomes more accurately without having to program them directly. With data mining tools, ML plays a significant role in disease prediction and healthcare organizations. ML techniques include support vector machine, decision trees, clustering techniques, and ANNs. ML intelligence technology is part of AI that enables computers to learn automatically based on their experiences without any need for user intervention [6]. It is a concept that is based on their machine, but it should only learn and adapt based on their experience. The real benefit of ML is that it allows computers to learn and assist themselves without the need for human intervention. The learning process begins with data collection, observation, experience, and examples and then moves on to identifying patterns in the data, making better decisions, and applying them to future predictions.

2.6 Machine learning methods

ML methods are classified as supervised or unsupervised learning.

2.6.1 Supervised learning methods

Supervised learning come under the type of ML, and it applies an earlier discovered technique that can be used for new labeled data, and machines predict the future output correctly. Here labeled data is nothing, some input data that is with the correct output. The data is trained using a labeled dataset in supervised learning, and the model learns about each piece of input. After the training phase is completed, the model is evaluated using testing data, and the result can be predicted.

2.6.2 Unsupervised machine learning algorithms

Unsupervised ML algorithm does not have any pre-assigned labels for the training data. To train the information is neither classified nor labeled. Unsupervised learning investigates how computers might predict a function with unlabeled data to determine a hidden pattern. Unsupervised learning algorithms must, as a result, first self-discover any naturally occurring patterns in the training data set. Usually, the semi-supervised learning method constantly improves the accuracy level. When we compare to supervised learning, unsupervised learning performs a more complex task. Unsupervised learning is not directly applied to classification problems or regression techniques because the input data is present, but there is no corresponding output data. Example: Assume that the unsupervised learning system has an input dataset including images of various cats and dogs. The algorithm is not ever trained on the given dataset; therefore, it does not know its features. The unsupervised learning algorithm's goal is to recognize visual elements on its own. This assignment will be completed using an unsupervised learning algorithm clustering the image collection into groups based on image similarities.

2.6.3 Reinforcement machine learning algorithms

Reinforcement learning is next type of ML learning approach for making a series of actions and engaging with its circumstances by producing activities and spotting errors. This function automatically invokes the ideal behavior within a specific environment to maximize performance. The agent learns to achieve a goal in an unpredictable, relatively dynamic environment. AI meets a game-like circumstance in reinforcement learning. To find a solution to the problem, the computer uses trial and error. AI is given rewards or penalties for its actions to accomplish what the programmer desires [7]. Despite the fact that the designer specifies the policies and rules the algorithm was not provided any advice or ideas for solving the game. Starting with completely random trials, it works way up to intricate tactics and superhuman abilities. In the case of inabilities, the model must decide how to finish the assignment to optimize the benefit. Using the power of search and a large number of trials, this algorithm is currently the best way to indicate to machine creativity. Unlike humans, AI can gain knowledge from large numbers of concurrent games if a reinforcement learning algorithm is run on a super computer. Large amounts of data can be examined using ML [7]. While it generally yields better and more precise results in trying to identify significant benefits or dangerous threats, completely training this could take additional effort as well as resources. Combining ML with AI and intellectual technologies can help it process massive amounts of data more efficiently.

2.7 Deep learning and its types

Deep learning is a subset of ML, and it facilitates in the automatic extraction of key features from raw data. Unlike traditional ML algorithms, which are sequential, deep learning algorithms are packaged in a ranking of increasing the complexity and abstraction. With the help of a superior learning algorithm, deep learning improved the predictive size of the computing devices. More profound education's computational roots are quickly grounded in traditional neural networks. Deep neural networks (DNNs), like conventional neural networks, are made up of artificial neurons organized into input, hidden, and output layers. When we compare to traditional neural networks, DNNs have more than one hidden layer. The DNN can learn multiple levels of features, and each level of an element corresponds to a specific group of abstraction. At the basic level, initial components are known, then embedded in deeper layers to build higher-level concepts [8]. Deep learning generates complex structures from massive datasets. The backpropagation technique instructs the computer on where to modify the system parameters used to calculate the interpretation for each layer based on the expression in the previous layer. In deep learning methods integrate ML with an application such as self-driving cars, drug discovery, diseases detection [9]. Nowadays, deep learning methods have given more attention in science, so the researcher attracts it.

2.8 Deep learning models

There are numerous deep learning methods available, including autoencoders, multi-layer perceptron, convolutional neural network (CNN, and recurrent neural networks).

2.8.1 Autoencoder

Autoencoder has come under the category of unsupervised neural network. Autoencoder is mainly used in complex high-dimensional data. It describes how to represent the data through dimensionality reduction, and it is used to reduce the sizes of the input into a more miniature representation. If we need original data, it can be reconstructed from compressed data. AI uses a wide range of technologies and techniques to solve computer system problems like a computer network, data compression, computer architecture, etc. In autoencoder uses ML to do this compression for us (Fig. 2.1).

Autoencoder is used in various applications such as image coloring, dimensionality reduction, watermark removal.

Figure 2.1 Autoencoders.

2.8.2 Types of autoencoders

2.8.2.1 Convolution autoencoders

The primary and simplistic autoencoder limits the number of the nodes in the hidden layers, thereby restricting the flow of information from the input to an output node. It penalizes the unimportant feature; thereby, only essential and relevant elements are passed input to output. On the other hand, the decoder reconstructs the data from the hidden layer to the production again; thereby, original data is obtained.

2.8.2.2 Sparse autoencoders

Sparse autoencoder offered an alternative way for generating an information bottleneck in which nodes reduction is not necessary at the buried layer. Instead, we build our loss function and fine-tune the layer's activation (Fig. 2.2).

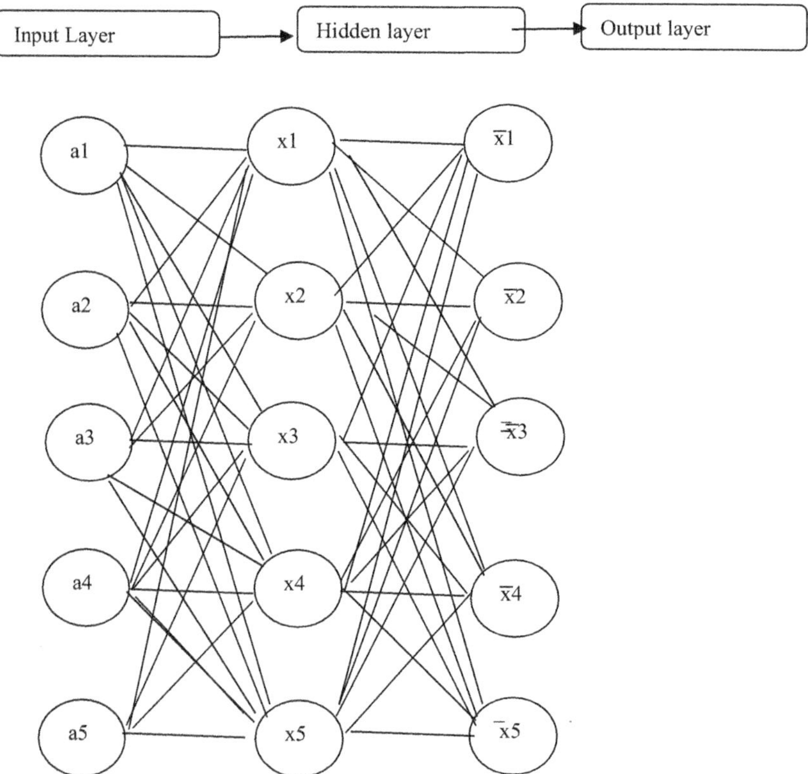

Figure 2.2 Sparse autoencoder.

2.8.2.3 Deep autoencoders

Deep autoencoder is the extension of the autoencoder. There are two layers in the deep autoencoder. The encoder contains four to five layers of hidden layer network that converts the raw inputs and applies filters based on the essential features. The decoder also contains the same number of hidden layers that retrieve the original information back.

2.9 Deep neural networks

Deep network is a Generative Adversarial Network (GAN) is as type of GAN that has a deep architecture of restricted Boltzmann. Each Restricted Boltzmann machine performs a non-linear transformation of the input vectors and provides the Output vectors. Deep belief networks can be used for supervised and unsupervised learning and can be used for feature extraction. The deep belief network can be used to generate an image, classify images, recognize video, motion capture, and natural language processing.

2.10 Artificial neural networks

Deep learning concept work on ANN where a group of nodes or units known as artificial neurons that functions similarly to a biological brain. The information is processed by neurons using a non-linear mapping approach. The neurons' role is to receive a signal from an artificial neuron, process it, and send it to other neurons. The movement was connected using real numbers at first, tying the neurons called edges; output was determined using a non-linear function of the sum of its inputs [10]. Some of the challenges experienced by non-linear methods are overcome with the help of these connections, complex equation problems are solved, and weight is constantly adjusted for an accurate result. The weight is increased or lowered based on the strength signal, and it can also be calculated using the threshold value. Neurons are separated into layers, and each neuron can respond to its inputs differently. The input layer is the first, while the output layer is the final. The signal travels from the input to the output layers. The information of the input node is numerical, and each piece of information represents an activation value; a number represents each node. As the level of activation has increased, so has the number. The values of activation are passed on to the following node [11]. Each node calculates weight and updates the value of the activation function. The activation function value applied to the neuron based on that output it decided whether to forward the signal or not. This process runs until it reaches the output node. The network employs the cost function to compute the gap between the estimated and calculated values. ANNs are used in various applications, including parallel processing, adaptive learning, fault tolerance, and data storage distribution, text classification and categorization, paraphrase. Forecast, image processing. ANN approaches work on biological instincts and helps to solve all physical, real-world problems. ANN techniques are used to solve problems in the power system, such as energy distribution, generation, and transmission. Given the constraints of a practical transmission and distribution system, the exact values of parameters can be determined. ANNs, for example, can calculate the values of capacitance, inductance, and resistance in a transmission line while accounting for a variety of factors such as environmental factors, unbalancing situations, and other potential difficulties. Resistance, capacitance, and inductance of a transmission line can also be given as inputs, with a combined, standardized, and the parameters' values accessible. In this way, the skin impact and similarity effect can be mitigated to some extent.

2.10.1 Artificial neural network characteristics

ANN does not require any system model knowledge. ANN works on this type of file when the data is incomplete or corrupt. ANN only works on the things for which they have been trained. Retrain the data if we want to complete any task. ANN will produce the result even if the input data is incomplete.

2.10.2 How artificial neural networks can be used in power systems

Although ANNs are related to biological instincts and carry out physical evaluations of real-world problems, problems in energy production, transmission, and distribution could be fed to the ANNs to find a suitable solution. Because of the constraints of practical transmission and distribution networks, the exact values of parameters can be defined. ANNs, for example, can calculate the values of capacitance, inductance and resistance in a power line while taking account for environmental parameters, unbalanced voltage conditions, and other possible problems. Resistance, capacitance, and inductance of a power line can also be provided as input data, and a merged, normalized value of the variables can be obtained. In this case, the skin effect and proximity effect can be significantly reduced.

2.10.3 Applications

ANN obtains inputs after generating results fast, ANN is always used in problems that require quick outcomes, such as real-time operations. The application of ANN to a power system problem involving the encoding of an unknown non–linear function is appropriate.

2.11 Convolution neural network

The CNN algorithm works on the images by applying the weight and bias on the input image, thereby differentiating between the images. The data pre-processing steps required in CNN are comparatively less while comparing it to other ML models (Fig. 2.3).

While filters are done by hand-engineered in the primary method, the CNN can do it by training itself. An image is made of pixels of different dimensions like 1×1, 3×3. When the image is of a smaller size, it can be flattened to a single extent. It can be used in a simpler feed-forward neural network, and also the prediction will be very average in the case of very complex images where the dimension is very high. The role of CNN is to reduce the appearance of the higher extent to lower size by picking up the critical feature in the images without losing the information of the images. To filter the images CNN use Kernel for example, a 5×5 dimension images will be converted to 3×3 in the first step by applying the filter with kernel 3. The kernel chooses 3 pixels of column and height and multiplies original images to pick the critical feature from the authentic images [12]. The next important term in CNN is a stride to select the number of pixels from the original images and jump to the next matrix based on the

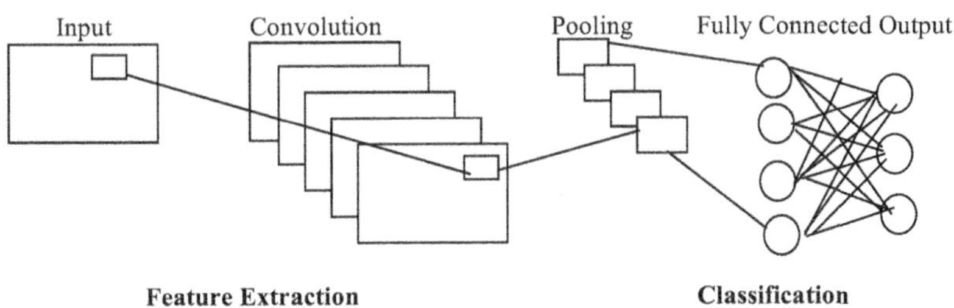

Figure 2.3 Convolutional neural network.

stride value chosen. CNN is a DNN. The initial layer captures the edges of the images and the deep network understands the information about the images accurately, similar to that of the human being. The next layer in the CNN is the pooling layer used to reduce the system's computational power by dimensionality reduction. Under pooling, there are two types (1) Max pooling and the other is average pooling. In Max pooling based on the kernel, the max of the matrix is taken to the next steps. Other pixels are filtered out as only max pixel will have important information whereas in the average pooling, the average of the matrix selected is taken forward to the next steps [13]. Among the two, max-pooling performs better in most situations as it eliminates the noise in the input images. Now a full CNN is created, and the last layer is to flatten the images, and the SoftMax activation function is applied to find the classification of the images. There is various form of CNN, and the important ones are highlighted in the following sections. Each type has a slight variation between them. Alex Net, LeNet, Google Net, GNet, ZFNet, RESNet.

2.12 Fuzzy system

Fuzzy logic system is responsible for controlling, which is heavily used in machine control and is based on fuzzy logic. The fuzzy system works in the same way that a mathematical system does, by storing the input value in a logical variable with a continuous value of 0 or 1. A fuzzy system is referred to as a human decision system because it can produce precise results from unique or even approximate knowledge / analysis. The fuzzy system is similar to human brain, and that we can use these principles to instruct machines to function like humans. Fuzzification improves expressive capacity, generality, and the ability to describe complex issues at a low or moderate cost of solution. Fuzzy logic allows for some ambiguity throughout an analysis [14]. The fuzzy system is used in many applications where the information is indecision. Logical reasoning problems, for example, are applied to computational models apart from representational inputs and outputs. It aids in the transition from numeric data to symbolic inputs and keeping control of the results.

2.12.1 Fuzzy logic usage in power systems

The components of a power grid can be designed using the fuzzy logic system. It can be used in all types of networks, from small legacy systems to large mainframes, and it also improves the power system's capability.

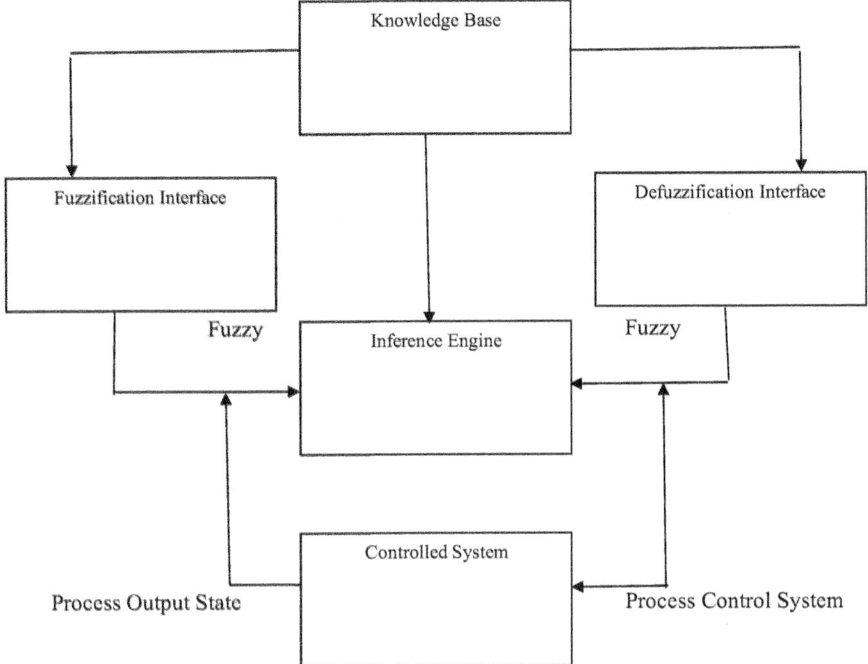

Figure 2.4 Power system.

The output of a power system can only be approximate or based on assumptions, but the production of a fuzzy system is stable, accurate, and error-free (Fig. 2.4).

2.12.2 Fuzzy controller

To design and control, a fuzzy controller generates a fuzzy code and provides broad input. The input can be implemented in either software or hardware, and it can range in size from significant to small circuits. The complex process is controlled by adaptive fuzzy.

2.12.3 Applications

Power system controls used in fuzzy system are the following:
- It is used in fault diagnosis.
- It helps in security assessment.
- It helps instability analysis and enhancement.
- it is used in reactive power planning and its control.

2.13 Expert systems in power system

As the name suggests, an expert system is a model that is expertized in a specific domain. It is similar to a specialist defined in a human job profile. The expert system is a model or program with more competence and knowledge in a specific field. Usually, it will be the form of a rule engine, compliance's, trees based on decision model or model trained to a particular data set and procedure and Frames [15]. These are not defined inside a structural program and will be part of an add-in that can be used in the different systems if applicable. The result from this expert system is dependable than a human being as it is trained and tested multiple times, and human beings cannot match such intellect.

2.13.1 Advantages

- One-time effort to build the model and thereby it will be very consistent.
- Will be saved as Ruleset or Engine, which can be documented for future validation purposes.
- Can be used as an Add-on in different solutions and models wherever the ruleset is applicable.

2.13.2 Disadvantage

The ruleset and model are developed for a specific domain and problem; this will fail if a new problem or additional charge is required.

2.13.3 Usage

It is used widely in the power system where the rules and decision making, knowledge based on the previous data and solve issues based on dataset judgment and reasoning. As the power system will have significant information related to transformation, units and this information need to be processed quickly, which will be impossible for the human to do effectively, efficiently, and repeatedly.

2.14 Expert systems in power systems

In a power system, the data will usually be related to power generation and ways of transformation from one grid to another or the power station and distribution of same to different industries, hospitals, and households. Hence, if a computer program or model can be developed to perform this operation effectively if slight modification is required, the program can be updated as needed. Further research is in the process of increasing the effectiveness of the system.

2.14.1 Artificial intelligence in genetic algorithm

Genetic Algorithm use heuristic search to solve or find the optimum solution for a given problem. These algorithms are better than other search algorithms as they have the intelligence to find patterns based on historical data [16]. Genetic algorithms are used in the place where complex problems are available, and impossible for human beings to find intuition from the dataset. These are used in a variety of real-world problems.

The basic terms used in the genetic algorithm are

- Population: Find the best set of solution for a given problem to resolve it.
- Chromosome: one of the best solutions from the subset identified.
- Gene: a member in the chromosome.
- Allele: Value assigned to a gene found in the chromosome.
- Fittest function: This function takes the solution available as input and outputs if the solution fits the provided problem.
- Genetic operation: This is one of the medical operation operations where a new offspring is generated from the parents. These offspring are genetically more efficient than the previous generation. The Fittest function plays an essential role in the output as the function should be very efficient and learn from all the available historical data.

2.14.2 Advantages of genetic algorithm

- Parallel execution and capabilities to find faster search.
- Find optimal solutions for different kinds of problems.
- The solution improves over time based on the historical data available for training.
- Continuous improvement.
- Derivative data or information is not required.

2.14.3 Steps in genetic algorithm

- Initialize the population

    The first step in the Genetic algorithm is to identify the population for the given problem. This population will contain all the solutions required or identified for the issue at hand.

- Fittest function

  As stated earlier, the fittest function is an essential step in the genetic algorithm. The search technique and solution for the problem are found by providing a score for each solution found for the provided issue. As always know, the higher the score, the higher the chance for the next step, reproduction.

- Selection of individuals

  Based on the score from step 2, the individual with higher scores is selected for the production of offspring. The offspring will be more efficient and will have all the characteristics of the parents from which it was created.

- Reproduction

  This is the actual steps where solution are found based on the parent solution or previous generation solution. It is achieved through two phases.

- Crossover

  The parents are matched with different combinations to generate the best offspring.

- Mutation

  New intelligence is added to the offspring so that the offspring is much more intelligent than the parents.

- Replace

  After the new offspring is found, the parents are replaced with offspring. The offspring, as stated, will have a higher score of fitness, intelligent decision making, etc.

- Terminate

  Based on the threshold set, the algorithm exists after the prescribed value is reached.

2.14.4 Applications in power system

- In a power system where wind turbine positional and optimization are required.
- In power system for routing and place selection to keep the grid, capacitor, and transmission system.
- Coordination between different thermal units.
- Filter design.
- Power loss minimize
- Distribution of power load to a different unit.

2.15 Genetic algorithms in power systems

The genetic algorithm is based on finding the best solution for the provided problem that can be used effectively in the power system. The performance of the production of power generation can be obtained. AI uses in power systems are the following:

- Power system involves hydrothermal, traffic management, load management, maintenance, scheduling, and power flow, which can be effectively solved through AI.
- Combining AI and expert systems can be used to plan electricity generation depending on demand, power system reliability, and other factors. Maintain optimal voltage for devices and home usage, the flow of power control, and control on the frequency.
- Location identification, device controlling and ideal size of device identification.
- Power restoration automation, identifying faults in the network, and security issues handling.
- Power distribution effectively uses a distribution system by identifying the demand for the power required in smart grids.
- Usage of solar photovoltaic power, turbine wind plan
- Using AI for power forecasting.

2.16 Conclusion

Conventional techniques used in the power system do not fulfill the existing problem and use in the current age. Owing to the traditional technique, the maintenance cost and operational cost are getting higher. The current new-age data-based AI system solves the majority of issues that exist in the power system. It takes full advantage of upcoming technology for generating electricity, effective distribution, and developing renewable energy to reduce costs. This chapter highlights various techniques in AI and its application in the power system to reduce the use of conventional methods, which will improve the power system and make informed decisions to minimize losses and improve efficiency.

References

[1] J.C. Bose, University of Science and Technology, YMCA, formerly YMCA University of Science and Technology and YMCA Institute of Engineering, is a state public university located in Faridabad, in the state of Haryana, India.

[2] W.R. Anis Ibrahim, M.M. Morcos, Artificial intelligence and advanced mathematical tools for power quality applications: a survey, Power Delivery, IEEE Transactions 17 (2) (2002) 668–673.

[3] M.A. Laughton, Artificial intelligence techniques in power systems, in: IEE Colloquium on Artificial Intelligence Techniques in Power Systems (Digest No: 1997/354), 1997, p. 1.

[4] E. Charniak, Introduction to Artificial Intelligence, Pearson Education, India, 1985.

[5] Q.Y. Sun, L.X. Yang, H.G. Zhang, Smart energy—applications and prospects of artificial intelligence technology in power system, Control and Decision 33 (5) (2018) 938–949.

[6] B. Babic, S. Gerke, T. Evgeniou, I.G. Cohen, Beware explanations from AI in health care, Science (New York, N.Y.) 373 (6552) (2021) 284–286.

[7] B.R. Kiran, I. Sobh, V. Talpaert, P. Mannion, A.A. Al Sallab, S. Yogamani, et al., Deep reinforcement learning for autonomous driving: a survey, IEEE Transactions on Intelligent Transportation Systems 23 (6) (2021).

[8] D. Machalek, T. Quah, K.M. Powell, A novel implicit hybrid machine learning model and its application for reinforcement learning, Computers & Chemical Engineering 155 (2021) 107496.

[9] Y. Bengio, Y. Lecun, G. Hinton, Deep learning for AI, Communications of the ACM 64 (7) (2021) 58–65.

[10] S. Jin, X. Zeng, F. Xia, W. Huang, X. Liu, Application of deep learning methods in biological networks, Briefings in Bioinformatics 22 (2) (2021) 1902–1917.

[11] M. Kubat, Artificial neural networks, An Introduction to Machine Learning, Springer, Cham, 2021, pp. 117–143.

[12] W. Samek, G. Montavon, S. Lapuschkin, C.J. Anders, K.R. Müller, Explaining deep neural networks and beyond: a review of methods and applications, Proceedings of the IEEE 109 (3) (2021) 247–278.

[13] A.E. Maxwell, T.A. Warner, L.A. Guillén, Accuracy assessment in convolutional neural network-based deep learning remote sensing studies—Part 1: Literature review, Remote Sensing 13 (13) (2021) 2450.

[14] S. Albawi, T.A. Mohammed, S. Al-Zawi. Understanding of a convolutional neural network, in: 2017 International Conference on Engineering and Technology (ICET), IEEE, 2017, 1–6.

[15] K.V. Shihabudheen, G.N. Pillai, Recent advances in neuro-fuzzy system: a survey, Knowledge-Based Systems 152 (2018) 136–162.

[16] L. Zhang, Y. Pan, X. Wu, M.J. Skibniewski, Expert systems, Artificial Intelligence in Construction Engineering and Management, Springer, Singapore, 2021, pp. 201–230.

CHAPTER 3

Recent developments of smart energy networks and challenges

Kannadhasan Suriyan[1], Nagarajan Ramalingam[2],
Manjushree Kumari Jayaraman[2] and Ramya Gunasekaran[2]
[1]Department of Electronics and Communication Engineering, Study World College of Engineering, Coimbatore, Tamil Nadu, India
[2]Department of Electrical and Electronics Engineering, Gnanamani College of Technology, Namakkal, Tamil Nadu, India

3.1 Introduction

The goal of this research is to uncover the issues and trends that are impacting how utilities utilize data and analytics to achieve their goals. 136 utilities from 24 nations responded to the survey. Among the data analytics application areas identified by the findings are energy forecasting, smart meter analytics, asset management/analytics, grid operation, customer segmentation, energy trading, credit and collection, call center analytics, and energy efficiency and demand response program engagement and marketing. Universities and industry will work together to train a growing number of energy data scientists, helping to close the skill gap in energy data analytics. Meanwhile, demand-side deregulation and smart metering advantages are hastening the birth of a wave of new firms. To make money, these start-ups collect and analyze smart meter data, then provide consumers and merchants with insights and value-added services. More information about industrial applications may be obtained from data analytics-based startups.

Analytics is the scientific process of translating data into insights to make better decisions. In this smart meter data analytics research, these three factors are investigated. Modern civilization is completely reliant on energy, and power outages have a detrimental effect on people's quality of life. As a result, efficient power supply control is required [1−5]. It has become a hot topic of debate due to the vast disparity between energy use and production. Cloud computing, the Internet of Things, Bluetooth, Zigbee, and Wi-Fi have all benefited from microcontroller-based wireless communication technology. Traditional huge electromechanical power

Smart Energy and Electric Power Systems
DOI: https://doi.org/10.1016/B978-0-323-91664-6.00004-8

meters have been replaced by the above-mentioned technologies. Smart metering gives consumers the ability to lower their costs and better control their energy use. Smart meters are accurate measuring devices with digital displays that can record the amount of energy used and automatically transmit that data to a meter data management system for storage and processing. The service provider hires employees to collect readings from these meters, which adds to the physical labor. Meter readers will have to go to each residence once a month to collect this data. This may lead to human mistakes in obtaining meter readings, as well as inaccuracies in calculating paid and outstanding bills. Utilities must have power, usage patterns, frequency consumption statistics, and a bidirectional link between meters at the end user and the data management system.

One of the most important requirements for the continuation of life's contents is electricity. It should be utilized with caution to get the most out of it. However, in our nation, there are numerous places where there is an excess of energy, while many others do not have access to it [6–10]. Our distribution rules are also partly to blame, since we are still unable to accurately predict our precise requirements, and power theft continues to be a problem. Consumers, on the other hand, are dissatisfied with the services provided by electricity providers. The majority of the time, they get complaints about statistical inaccuracies on monthly invoices. This allows us to monitor the meter and determine whether or not a problem exists. A circular metal strip spins in the preceding meter, and we compute the consumption based on that revolution. However, our meter is based on a pulse that is produced based on consumption, and we previously attached an android board to monitor the pulse, and a bill is issued based on the pulse. We want to get monthly energy consumption from a distant site straight to a central office with the assistance of this project. We can decrease the amount of human labor required to record meter readings, which are now recorded by visiting each house individually.

A line man serves an important role in gathering and handling field data in conventional monitoring, that is, human labor. However, as the size of consumption regions grows, this kind of manual activity is seen as time consuming and labor demanding. Energy consumption is anticipated to rebound as Malaysia's economy recovers from the financial and economic crisis. Energy producing capacity (which grows in response to energy demand) has risen by almost 20% in the past 3 years, from 13,000 MW in 2000 to 15,500 MW in 2003. By 2010, the energy producing capacity is projected to rise to 22,000 MW. Energy supply infrastructure will need to be constantly improved to satisfy rising demand while also being highly capital demanding. As a result, natural resources will be under severe strain, especially in emerging nations like Malaysia. At the same time, it is obvious that present growth, resource consumption, and environmental degradation trends cannot continue forever [11−15]. The sharp rise in local energy statistics demonstrates that demand cannot be met on the basis of supply alone, since energy resources will be one of the main problems in the future, even more so than they are today for certain nations, particularly emerging countries.

3.2 Smart energy

Both short- and long-term management of available energy resources are required. One of the frequent issues that adds to the current challenges is peak electricity in the afternoon. Meanwhile, the rising need for new power plants in a scenario where generation is already heavily reliant on coal imports is a significant problem that must be addressed. Power plant generation is closely linked to peak energy demand, and resolving this problem may result in a decrease in that demand, at least for the time being; that is, postponing the urgent construction of a power plant. An examination of local energy data reveals that thermal plants, which rely mostly on imported coal, provide the majority of power production. This article is primarily concerned with the development of a new intelligent power monitoring system is shown in Fig. 3.1. The designed embedded system package includes an advanced control technique that consists of

Figure 3.1 Smart energy.

two parts: a slave part that is controlled by a high-tech controller and a "master part" that is dedicated to a specific task that may require extremely powerful processors with some level of extensibility or programmability. In this paper, we will look at real-time embedded systems that are utilized in a communication-intensive environment and have certain unique characteristics that should be addressed while creating new applications. When we talk about real-time constraints, we imply that a real-time system must adhere to specified (bounded) response-time restrictions, or its accuracy would be jeopardized, perhaps resulting in failure. The reaction time is just as essential as the accuracy of the outputs in these systems. A real-time system does not need to be quick; all it has to do is provide accurate answers within a certain amount of time.

A real-time embedded system often monitors the environment in which it is placed, and if it fails to react to a request in a timely manner, the consequences may be catastrophic. The slave portion, which is placed on the "user location," has a 32-bit MCU that integrates sensing devices, UN limited-range GSM modem modules, and a user display section. This system would be able to predict the time as well as the location to show on the LCD at each of the user's specified places. The monitor and IDE components are included in the software section. To monitor and maintain the system, a graphical user interface (GUI) and software administration packages are also being created. For the advanced controller unit, MPLAB was chosen as the IDE and C18 as the compiler; therefore, both hardware and software solutions are combined in the proposed system to make it the most dependable and user pleasant. This necessitated the use of core software to design the setup and create a copy of the actual network. The detected inputs are sent to a high-tech chip, which is responsible for producing and displaying the outputs on a continual basis. A wireless link connects each device to the master node, which houses the network operating center (NOC) and automated monitoring server. Predefined variable energy packages are the method used to reduce peak energy usage, encouraging the user to alter his consumption patterns. The method is proven to be a real-world application of embedded devices for power system network monitoring.

Several electrical firms may no longer be the primary supplier for both industrial and residential applications. Price disparities were likely to affect customer decisions. Due to cost and scale advantages, the bigger power plants have produced the majority of energy. They've produced and transported electric power over great distances and at various voltage levels. These hierarchical processes have been used to regulate, produce, and maintain the distribution system. Traditionally, it has been assumed that electrical energy flows continuously from substations to the feeders' ends. However, when a distributed generator is used in a deregulated environment, the power flow is reversed, resulting in problematic voltage profiles in the distribution system. As a result, changes to the distribution system's planning and operating methods are required. a different word Grid computing is a critical component. To accomplish a goal on a big scale resource, it varies from conventional methods. As a result, the development of grid computing relies on programs to take use of the technology.

Grid advancements have resulted in the introduction of wireless sensor networks, which have become a constant development in smart grid technology. These sensors make a provincial choice, and the data is gathered to create a complete picture of their surroundings. Nodes collaborate in sampling, data gathering, and status monitoring because of the strength of sensors. In addition, the Grid concept is being used to build an expandable multimedia server. This technology's goal is to help virtual organizations coordinate their many resources. Grid systems nowadays are based on the X.509 public key architecture. Security management is a hot topic these days. As a result, smart card technology is used to keep data safe from unwanted users.

The term "distributed computing" refers to a collection of technologies. These computer systems provide support for the future of power and energy distribution systems. An automated meter reading system is also implemented to ensure that consumers' power usage readings are accurate. It relies on a variety of communication methods to transmit data across long distances. Because of the information and communication structure, the implementation of a smart grid is dependent on consistent data transfer. As a result, power line communication technology has been deployed to expand the power communication network to consumers.

Demand side management is another key technique that guarantees the power system's stability and accuracy. There are a variety of demand side management methods that may be used to residential, commercial, and industrial energy management. There are methods for dealing with uncertainties that are shown in the context of developing smart grid problems. There are many artificial neural network forecasting methods that imitate the human brain to achieve regularities and generate generalized findings. Further, the US Department of Energy defines demand response as changes in electrical power consumption by users as a result of their use techniques changing energy costs over time. In this article, a systematic and comparative examination of several smart grid technologies in terms of electrical power security and development is addressed. So, in preparation of the arrival of machine to machine communication, in which all appliances may be wirelessly enabled through Zigbee or Bluetooth, we've constructed a small IoT prototype system that provides temperature, humidity, and light intensity data using a Hall and Light intensity sensor. The data is sent into an Arduino microcontroller, which then sends it wirelessly to Raspberry Pi3 edge processors. The recommended system includes a Raspberry Pi3, an Arduino microcontroller, a Wi-Fi shield, as well as modules like a Hall sensor, a light intensity sensor, and an ambient temperature sensor. Based on the humidity, temperature, and lighting conditions in the surroundings, the Arduino microcontroller will alter the appliance's settings, such as fan speed and light intensity, resulting in decreased energy usage. These current amounts are recorded and sent over Wi-Fi to a Raspberry Pi3, which calculates the power use and presents a graph, which is then uploaded to a cloud server. As a consequence, depending on the climate, the system has achieved energy consumption in appliance usage, such as fan and light. This gives the user real-time information on appliance power use.

3.3 Challenges in smart energy

One of the experiments presented was a Raspberry Pi-based IoT-based Automated Temperature and Humidity Monitoring and Control System. The Pi detects and transmits temperature and humidity data. This research, on the other hand, resulted in the creation of a prototype for automated temperature and humidity management that is both functional and feasible. A user-friendly GUI has also been designed for an IoT-based Smart Home Control and Monitoring System that can be accessed from any device with an internet connection anywhere in the globe. A Smart Home Monitoring prototype was constructed utilizing an Android phone and wireless sensor devices in addition to the studies mentioned above. This technique monitors the electrical power consumption characteristics at the socket outlet in real time. This system continuously monitors the voltage, current, and temperature of socket outlets in each room and feeds the information to the system, which computes the threshold violation and alerts the user before the circuit breaker trips or a fire occurs.

In addition, research was done on creating an Automatic Lighting and Control System for Classrooms to save energy. They've also given the system mobility and remote command execution through Bluetooth, allowing it to control lights via voice commands. To regulate energy at the appliance level, the Energy Management System for Smart Home was developed. As a consequence, a Smart Home Energy Management System Architecture was developed to assist in this endeavor. The energy consumption of home appliances is controlled by sensors in this system. Solar energy is also used as a backup supply in cases when resources must be relocated due to weather conditions. The PC server collects and compares energy data from many home servers to give statistical analysis data. In addition, for Myanmar's rural regions, an IoT-based Home Energy Management system has been developed. In this research, the demand for electricity was projected, and suitable methods were established to meet the need. To address energy needs, non-conventional energy sources such as solar and thermal may be utilized.

Under the existing billing approach, distribution companies are unable to keep track of altering maximum demand. Even if bills are paid on time, the client may experience problems such as late invoicing for payments that have already been made, as well as poor power supply and quality. All of these issues may be solved by keeping track of the consumer's load on a regular basis, which will allow for correct invoicing, tracking maximum demand, and detecting threshold values. All of these aspects must be considered when developing an effective energy billing system. The current project, "IoT Based Smart Energy Meter," tackles issues that customers and distribution businesses confront. The focus of the article is on smart energy meters, which make use of embedded system characteristics such as a mix of hardware and software to achieve required functionality. The article compares Arduino and other controllers, as well as the use of GSM and Wi-Fi modems to establish the idea of "smart." Consumers and service providers will both get the utilized energy reading with the corresponding amount through GSM modem. Consumers will also receive notice in the form of SMS via GSM when they are close to exceed their threshold value that they have established. The user may also monitor his consumption readings and adjust the threshold value through a website using a Wi-Fi connection.

This method allows the energy department to check meter readings on a monthly basis without having to send someone to each home. This may be accomplished via the use of an Arduino device that constantly monitors and stores energy meter readings in a non-volatile memory location. This technology constantly records the readings, and the customer may request that the live meter reading be shown on a website. This method may also be used to turn off the house's electricity supply as necessary. An energy meter, also known as a watt-hour meter, is a device that monitors the quantity of electrical energy used by users. Utilities is one of the electrical departments that installs these devices in various locations such as houses, businesses, organizations, and commercial buildings to charge for the energy used by loads such as lights, fans, refrigerators, and other household equipment.

The energy meter calculates the product of the fast voltage and currents and provides instantaneous power. The overall quantity of energy utilized at that time is calculated by adding up this power over a period of time. On a daily basis, the population is growing. As a consequence, both residential and commercial regions need a substantial amount of current. Various energy-saving solutions for home electric meter devices have previously been developed. In the past, the energy meter gadget was installed inside the consumers' homes. The consumption rate was recorded and put into the system by humans. The operator was highly important in this system. The operator drew it to the users' locations to gather data. It is a time-consuming and exhausting process. The recommended method monitors the current consumption rate automatically to avoid this issue. This system uses IoT to gather the current usage data and provides the user an alert message. As a result of the IoT approach, a precise result as well as financial advantage are produced. Since its conception, this technology has advanced to the point that it is now in use. Electricity is fundamental to everyone's life. People in today's civilization would perish if they didn't have access to energy as shown in Fig. 3.2.

This proposed system captures information about the user's location and applies it to the billing process. The calculated amount is based on the measured value, which is recorded on the server. This technology correctly measures current consumption units. This technology eliminates the need for human involvement. On the LCD panel, the suggested system shows real-time current consumption statistics. A website may be used to verify the number of used units and provide a threshold unit level using a Wi-Fi modem. Electronic intelligence—the root of this transformation—has already taken over economic sectors from manufacturing to commerce. Since its inception, information technology has been used in electrical grid command centers, and it is now beginning to inject

Figure 3.2 Challenges in smart energy.

electronic intelligence throughout the system. The conventional grid is in the early phases of transition to a smart energy network, driven by the recent development of cheap processing power and low-cost bandwidth. For electrical power ratepayers, this means lower electricity costs. They'll also save money thanks to smart technology that boost energy efficiency in homes and workplaces. Everyone concerned about power reliability will benefit from the smart network's increased capability to recover from outages, resulting in fewer blackouts and brownouts. Smart energy reduces air pollution and greenhouse gas emissions, which is good for the environment. The ability of information technology to improve grid operations is critical to smart energy. Electrical devices will be controlled by intelligent software agents that communicate with the network about their operational state and requirements, gather data on pricing and grid conditions, and react in ways that benefit their owners and the grid. The grid will evolve from a centralized control system to a collaborative network similar to biological systems as a result of millions of smart agents' constant interactions and transactions. The smart energy network is beginning to gain traction, and it will reach critical mass within the next decade.

The smart energy network's information technology foundation is communications and control systems, which provide two fundamentally new capabilities:

1. The capacity to control electrical power demand down to the home level with precision.
2. The ability to connect a large number of small-scale distributed energy production and storage units to form a network.

From those two fundamental skills spring a slew of new and improved capabilities, resulting in a flood of new electrical power gadgets, services, markets, and players. This is expected to bring significant changes to the electrical system, as well as significant advantages and possible instabilities. If society wants to navigate through this developing information-technology-rich electrical network, avoid traps, and reap the full spectrum of advantages in the shortest period feasible, a complete strategy is required.

3.4 Conclusion

Smart meters can read and securely communicate real-time energy usage data such as voltage, phase angle, and frequency. Because smart meters can transfer data in both ways, data on energy delivered back to the power system from customer houses may be collected. A smart meter system consists of a smart meter, communication infrastructure, and control devices. Both locally and remotely, smart meters can communicate and carry out control commands. Smart meters may be used to monitor and control all of the customer's household appliances and gadgets. They may also gather diagnostic data on the distribution grid and household appliances, as well as interact with other meters within their range. They may monitor grid power usage, enable decentralized generating and energy storage systems, and charge customers appropriately. Data include a unique meter identification, a timestamp for the data, and the amount of energy used. Smart meters may be configured to solely charge for electricity used from the utility grid, leaving power consumed from distributed production sources or storage devices owned by consumers unaccounted for. Smart meters may set a maximum energy usage restriction and remotely disconnect or reconnect electricity service to any consumer.

References

[1] D. Sharayu, H. Dhanawade, Smart electricity meter using LoRa module, 4th International Journal of Advanced Research, Ideas and Innovations in Technology 5 (4) (2019) 146−148.
[2] C. Yao, H. Saputra, L. Meng, et al., Secure smart metering based on LoRa technology, in: 4th International Conference on Identity, Security, and Behavior Analysis (ISBA): Singapore, Proceedings 1−8, Research Collection School of Information Systems, 2018.

[3] F. Abate, M. Carratu, C. Liguori, M. Ferro, Smart meter for IoT, International Instrumentation and Measurement Technology Conference (I2MTC), IEEE, Houston, TX, 2018.

[4] B. Stephan, L. Szwec, Smart meter with smartphones: user − centered design of a mobile application in the context of energy efficiency, International Conference of Design, User Experience, and Usability (2013) 631−640.

[5] P. Loganthurai, M. Shalini, A. Vanmathi, M. Veeralakshmi, V. Vivitha, Smart energy meter billing using GSM with warning system, in: International Conference on Intelligent Technologies in Control, Optimization and Signal Processing (INCOS), Srivilliputhur, 2017.

[6] W. Gunawan, G. Permata Saktiaji, I. Ibrahim, et al., Techno economic analysis of smart meter reading implementation in PLN Bali using LoRa technology, in: International Conference on Broadband Communication, Wireless Sensors and Powering (BCWSP), Jakarta, Indonesia, 2017.

[7] F. Shrouf, J. Ordieres, G. Miragliotta, Smart factories in Industry 4.0: a review of the concept and of energy management approached in production based on the Internet of Things paradigm9−12 December Proceedings of the 2014 IEEE International Conference on Industrial Engineering and Engineering Management (IEEM), IEEE, Selangor Darul Ehsan, Malaysia, 2014pp. 697−701.

[8] D. Bandyopadhyay, J. Sen, Internet of things: applications and challenges in technology and standardization, Wireless Personal Communications 58 (2011) 49−69.

[9] International Energy Agency (IEA), Global energy & CO_2 status report, 2019. Available online: https://www.iea.org/geco/ (accessed on 27 September 2019).

[10] M. Lavanya, P. Muthukannan, Y.S.S. Bhargav, V. Suresh, IoT based automated temperature and humidity monitoring and control, Journal of Chemical and Pharmaceutical Sciences, ISSN: 0974-2115.

[11] A. Rajurkar, O. Shinde, V. Shinde, B. Waghmode, Smart home control and monitor system using power of IoT's, International Journal of Advanced Research in Computer and Communication Engineering 5 (5) (2016).

[12] S. Sankaranarayanana, A. Thien Wanb, A. Harayani Pusac, Smart home monitoring using android and wireless sensors, International Journal of Engineering and Manufacturing 2 (2014) 12−30.

[13] S. Suresh, H.N.S. Anusha, T. Rajath, P. Soundarya, V. Prathyusha, Automatic lighting and control system for classroom, in: Proceedings of 2016 IEEE International Conference on ICT in Business, Industry and Government, Indore, Madhyapradesh, 2016, pp. 1−6.

[14] S. Kshirsagar, D.E. Upasani, Energy management system for smart home, International Research Journal of Engineering and Technology (IRJET) 3 (6) (2016).

[15] K. Vinay Sagar, S. Kusuma, Home automation using internet of things, International Research Journal of Engineering and Technology (IRJET) 2 (3) (2015).

CHAPTER 4

A smart and efficient IoT-AI and ML-based multifunctional system for multilevel power distribution management

Lakshmi Kanthan Narayanan[1], Priyanga Subbiah[3],
S.A. Sahaaya Arul Mary[2], R. Rengaraj Alias Muralidharan[4],
Iswarya Gururajan[2], Ranjani Sampathkumar[2] and
Arun Prasad Baskaran[2]

[1]Department of Computer Science and Engineering, Quannta School of Software Engineering, Villupuram, Tamil Nadu, India
[2]Department of Computer Science and Engineering, Saranathan College of Engineering, Trichy, Tamil Nadu, India
[3]Department of Networking and Communication, SRMIST, Chennai, Tamil Nadu, India
[4]Department of Information Technology, Saranathan College of Engineering, Trichy, Tamil Nadu, India

4.1 Introduction

Friction is the least accurate of the six power processes. When a fabric scrapes against another object, friction electricity happens. As a result of the rubbing process, the object acquires charge and now possesses an electrical charge. Positive and negative electrical charges are the two most common forms. The positive and negative charge forms attract the positive and negative charge types, but the same charge type repels itself. For example, when contrary charges come into touch, they repel one other, and when same charges come into contact, they attract each other. Van de Graaff generators, which are used to generate large voltages to test the dielectric strength of insulating materials, are most commonly used for static electricity. Other uses include electrostatic painting and sandpaper manufacturing. The coarse grains create a negative charge when they pass across the negative plate. The coarse grains are pulled to the positive plate despite the fact typically opposing charges attract, allowing particles to be buried in the cement due to their high impact velocity.

The voltage established by the junction of two metallic materials is a function of temperature, as Thomas Seebeck discovered in 1821. When a

Smart Energy and Electric Power Systems
DOI: https://doi.org/10.1016/B978-0-323-91664-6.00008-5

closed circuit consists of electrodes of two different metals and one junction of the two different metals is at a higher temperature than the other, an electromotive force of a certain polarity is generated. Electrons flow from the hot to the cold connections over the iron in the case of copper and iron, for example. Electrons move from iron to copper at the required temperature, while electrons flow from copper to iron at the surface of the cathode. The Seebeck effect is a property of electromotive force generation. This phenomena is included in the most widely used temperature measurement method.

The SCADA System represents the gift of electricity that science has bestowed on humanity. It has also become a necessary part of modern living, making life difficult to comprehend without it. All of this gives them a feeling of security. Massive machinery in factories are powered by electricity. Power is used to make food, clothing, paper, and a variety of other items.

All appliances, amusement, illumination, and, of course, all electronics, beginning with your home, require power. Electric trains, planes, and even some automobiles require the use of power stations. Schools, medical facilities such as hospitals, and retail outlets all rely on electricity to work effectively. Electricity allows for the use of medical technology. It allows these medical facilities to work more efficiently.

A range of measures are used by diverse stakeholders in contexts ranging from local to regional levels to help electric transportation infrastructure and outcome. Many of these models depict the electric system as it was in the past, which is still the case in major parts. With the increase of generation sources actually connected to distribution systems, rising amounts of variable renewables and inverter-based power, and efforts to electrify transportation and building energy consumption to minimize carbon emissions, the electricity system is evolving dramatically. These transitions, which are expected to accelerate and deepen in the coming years, pose a danger to conventional design models, assumptions, and processes. This workshop will examine some of the most recent models used in electricity system planning, as well as modeling and research needs and prospective approaches to assist in the planning of electric power's future.

4.1.1 Electricity meters, their varieties, and their history

Electricity meters are electrical equipment that use readings to calculate and show the amount of energy utilized. Since the late 1800s, traditional meters have been in use [1]. They send messages between electrical components in a computerized setup for both power production and transmission. Most traditional energy meters use aluminum disks to measure power use. LTDay's power meters are not without problems, despite the fact that they are digital.

In this chapter, the sections are organized in the following manner. Section 4.2 discusses about the integration of Internet of things (IoT)-artificial intelligence (AI) and machine learning (ML) toward power distribution and health monitoring systems. Section 4.3 discusses in details about the grid–assisted technology. Transformer health monitoring from traditional till modern techniques is elaborated in Section 4.4. Section 4.5 explains about the forecast models toward consumer demand and stock analysis. In Section 4.6, the transmission and distribution health monitoring systems has been discussed. Section 4.7 discusses about the IoT based smart architecture for power distribution.

4.2 Integration of Internet of things-artificial intelligence and machine learning toward power distribution and health monitoring systems

Transformers are used exclusively in the transmission and distribution of electricity. The task of dispersing the transformer in recharging settings (as specified on its nameplate) ensures a long service life [2]. However, when they are subjected to extreme loads, heating, or low- or high–voltage/ current, their life is significantly shortened, resulting in unexpected disappointments. Excessive and insufficient transformer cooling is the primary source of dissatisfaction. Because the transformer is made up of many interconnected components, all of them should be inspected on a regular basis to keep the transformer in good working order [3].

The concept proposed here is an introduction of the continuous transformer health monitoring system utilizing IoT. Four sensors such as level sensor, gas sensor, temperature sensor, and current sensor were involved. A power supply is used to operate microcontroller PIC16F877A. Once the data can be read from the LCD display and at the same time these values are transmitted to the IoT module through the UART cable and the IoT module sends the data to the user on given IP address as per program. This information sufficiently supports the transformer to avoid its sudden failure. For any abnormal conditions in the load side, we can control by switching off the relay and if the temperature rise above the threshold value, the cooling system automatically will be ON until the temperature value return to the threshold value.

Collusion detection criteria square measure computed for the equilibrium and their peripheral points. Simulation results show that the accuracy of the used machines is suitable in collusion detection. The impact of every criterion on the collusion detection accuracy has been evaluated and also the rate and Celtic deity indices play additional vital role in collusion classification performance than different criteria. The results show the collusion of two or more additional corporations for a protracted period of time, the collusion is valid. However, if the results show the collusion happens solely in associate degree hour, it can not be thought-about as collusion.

The load shifting is attended since the reinforcement learning (RL) algorithmic software gives a battery action because the output is in step with it. When the grid synchronises with the battery level, power is distributed to the highest-priority hundreds in the most efficient way possible. Within the Load shifting algorithmic program, a general algorithmic program is used to transfer the masses between the most grid and the battery in response to the battery action given, taking into account the user's pre-prioritized load list. RL is a type of ML that focuses on how software system agents should behave in a particular environment to maximize the concept of cumulative reward. An intelligent Energy Management System for a Residential Building is demonstrated using ML techniques. For the shifting hundreds between normal grid superpowers and non-conventionally positioned domestically energized battery storage electrified by the top star, relevant Battery optimization strategies have been taken with the use of ANN correlated anticipated solar energy generation and other characteristics. ANN and supervised learning are used to predict star PV generation. RL strategies are applied to optimize battery usage. The alphabetic character acquisition agent is being recruited for future work as a development technique.

With the rising popularity of renewable energy within the facility and the growing number of people, energy management has become troublesome due to the variability of generation capacity and the offer from renewable energy resources. Correct load forecasting, as well as PV electricity, can aid in meeting the provision demand balance, resulting in cost-effective home grid energy management. Predicting load demand and renewable generation, such as PV Power Output, is difficult due to the large influence of variables like climate and time on each variable. A combination of two factors will induce variation in load demand and PV production. The SVM (support vector machine) model was used to forecast load demand. SVR is a subset of the SVM model in which the distance between all data points and the plane is lessened by finding a regression plane. Multi-Layer Perception (MLP) commenced with a 6-layer model (1 input layer, four hidden layers, and first output layer) and adjusted the output dimension using the Keras framework.

4.3 Grid-assisted technology

"Smart grid" is the name given to a recent advancement in electricity grid technology. The electrical load variability of household gadgets has caused recent power infrastructure to become weak. Electrical and energy management systems are growing increasingly susceptible as the world's population grows. As the grid's population expands, so does the number of people on it.

The term "smart grid" refers to systems that enhances grid efficiency by remotely regulating and improving dependability, as well as recognizing consumption and communicating data (in real time) to customers and suppliers [4]. Smart Grids make use of automated sensors.

These detectors can identify rolling blackouts and preserve power lines from overheating by transmitting measured data to utility companies. It has a self-curing characteristic. Smart Panels are based on the Smart Grid concept. By 2030, its installations are anticipated to cut carbon emissions by 5% per year, and it has the potential to have a larger impact on environmental benefits. For long-term development and the construction of the new grid infrastructure, several jurisdictions have proposed smart grids.

4.3.1 Intelligent metering

A smart meter is an energy meter that measures electrical energy in kilowatt-hours (kWh) in a way that is environmentally friendly (kWh). It's a technology that immediately benefits those who wish to save money on their electricity bills. They are part of the Advanced Meter Infrastructure division and are in charge of sending meter readings to the energy provider automatically.

Precision meter readings will be provided as a result of the Smart Meter's adoption of firm characteristics. They keep track of how much they eat on a regular or less-than-hourly basis. Non-volatile data storage, remote connect or disconnect capabilities, a verification phase, and two-way communication are all features of a Smart Meter. They send the acquired data to the centralized meter via remote reporting. The Smart

Meter's functionality is monitored by this central meter. Smart metering enhances the administration and control of the electrical system from an operational standpoint.

4.3.2 Technology concerns

There are several methods for communicating between the meter and the household appliances. To connect household appliances to the meter, you can utilize a dedicated channel, a wireless connection, web-based communication, or power-line communication. By connecting the meter to the datacenter, the secure scenario may be maintained. When Smart Meters are linked to mobile phones, the true energy consumed by a gadget may be determined regardless of whether it is turned on or off, plugged in or not. It provides an overview of Demand Management deployments, features, and implementations in the Netherlands.

The deployment of one or more Smart Meters that are continuously monitored and send data to the customer comprises smart metering. Smart meters give customers with a variety of benefits, including safe, secure, and economical electricity.

The commercial implications of smart meters in terms of individual and corporate structure has been explored. In contrasted to mechanical meters, the advantages and usefulness of Smart Meters are discussed. The researchers actually want to know what the hypothesis is for the proposed questions in the study paper. To create an energy-efficient society, customers must be aware of how and where energy they utilize. As a result, this research presents a number of feedbacks for reducing the energy consumption and streamlining operations.

4.4 Transformer health monitoring from traditional till modern techniques

Power transformers (PTs) are responsible for a large amount of power grid maintenance [5] and wealth creation (PTs). Because of the privatization of the energy market, there are now a range of operation schedules for conventional power plants, which can lead to more difficult PT operating conditions, shorter maintenance cycles, and more complicated obligations. Early detection of expected problems in PT improves equipment dependability while reducing revenue loss from repair downtime, resulting in a win–win situation for the industry [6].

The randomness (E-PCA-C) methodology uses the weighted PCA method to build clusters from the collected PC1 and PC2. The entropy-clustering (E-C) method uses a weighted KMA to find clusters that can be used with the chosen gases and weights. Implementing both strategies has the purpose of learning about the advantages of lowering issue complexity and focusing on the most important aspects. After the two parameters given above, it is still unable to determine the status of each cluster's PT, which is a disadvantage when compared to supervised systems. Engineers utilize engineering procedures at the "Engineering interpretation" step to tackle this challenge. KGA5 is used to determine whether a cluster is normal or failing, as well as to assign a severity rating to the latter. The method is used to compare "normal" and "failure" circumstances, as well as rate each cluster (conditions 1—4) from best to worst. The total dissolved combustible gas (TDCG) indicator is used to accomplish this, and it is computed as follows:

$$TDCG = \sum_{g=1}^{G} x_i^g.$$

EMD generates [4] a collection of IMFs from current signals Ia, Ib, and Ic (for three phases). The number of IMFs is then reduced by the relief attribute analyzer to a more tolerable level. To find inaccuracies, this smaller set of IMFs is used to train NN. Then, on a new set of data samples, NN is put to the test. Finally, we validate the ANN using a second defect data set (which we acquired after training and testing).

Transmission lines [7] (TLs) are responsible for 80% of power system problems because they are exposed to a variety of environmental factors as well as animal and human interaction. Short circuit faults account for the bulk of the problems encountered by the TL. A three phase TL's voltage and current signals vary from their reference values during a fault occurrence, resulting in disastrous effects if not cleared in time. As a result, fault analysis has become a popular study topic among power engineers.

4.5 Forecast models toward consumer demand and stock analysis

Within the broader energy framework [6], the electric power sector is one of the most essential components. Because reliable consumption data from individually metered consumers is available, demand studies and load projections for this sector have a better foundation than those for other energy sources. However, such data may not always be available in a convenient manner for analytical purposes. These give a more extensive and accurate foundation for future estimates than the other sorts of data that are typically used to support non-electric energy sources. Electricity demand forecasting errors, on the other hand, are usually more serious due to long lead times for new equipment installations, high initial capital costs, and the difficulty of replacing electricity with other energy sources, with the exception of cooking, heating, and domestic lighting, which require different appliances. Forecasts that are too low may result in expensive, short-term solutions to capacity shortages, such as the building of high-cost generating units (diesels or gas turbines), or rationing or forced supply interruptions. Lack of capacity has effectively moved industrial expansion and economic growth in some locations. Fluctuation, on the other hand, will cause scarce investment money and other resources to be squandered. As a result, load forecasting is an important aspect of the electric system planning process.

The demand drivers [8] for the remaining, ubiquitous sectors must be carefully picked, as they vary from case to case, and the level of disaggregation chosen will be determined by the data as well as the projection's ultimate goal. Demand forecasting is a dynamic process in the sense that it must be repeated frequently as new information and analysis tools become available. In this process of ongoing updating, developing a solid data bank is a useful tool.

4.5.1 Residential sector

The total number of households [9] (or population). Weighted real family income by income distribution. Electricity costs. Costs of connection. Service accessibility (urban, rural). Service dependability. Electricity-using fixtures and appliances cost and availability Consumer credit availability and cost. Alternative energy source costs (kerosene, LPG, natural gas, charcoal, firewood, coal). Days with temperatures that are below or above the range of adequate comfort.

4.5.2 Sector of commerce

Non-subsistence commercial sector sales or value-added. Electricity prices. Connection fees. The cost of electricity-powered appliances against those powered by alternative fuels for air conditioning. Alternative fuel expenses for air conditioning, including appliance costs. Service dependability. Different sorts of commercial establishments have different working hours.

4.5.3 Sector of industry

Electricity price (demand and energy charges by service type). Industry type and electric intensity. The degree of market power held by each industry. Alternative energy sources' relative cost. Supply reliability. Auto generation total costs comparison accessibility.

4.5.4 Agricultural sector

Electricity costs. Accessibility. Service dependability. Alternative energy system supply costs, such as diesel-powered irrigation pumps and milling and processing equipment, kerosene for lighting, and so on. Auto generation total costs comparison.

4.5.5 The government sector

Electricity costs. Municipal government revenue per capita. The number of public schools and their size. Expenditures on the armed forces and law enforcement (i.e., size of forces). Different types of public water supplies (e.g., deep-well pumps).

4.6 Forecast models toward consumer demand and stock analysis

The distribution transformer, which connects the generator to the clients, is one of the most significant components on the distribution side. The failure of a distribution transformer might occur if it is not monitored on a regular basis. Faults include overcurrent, overvoltage, a rise in the core of the transformer, and a surge or fall in the oil level.

The proposed system is an IoT-based distribution transformer health monitoring system, which includes devices that monitor and record the transformer's characteristics. It's necessary to keep track of voltage waveform, load power, temperature, and insulation oil qualities.

The Arduino UNO is a microcontroller that may be customized in a variety of ways to match the needs of the user. The Arduino UNO has 32 kB of flash memory and 2 kB of SRAM, making it more powerful than other low-cost computers. The Arduino microcontroller turns digital data into a string that the network connectivity module can read and transfer in the proposed system. The microcontroller's reset pin will format the entire chip, allowing us to upgrade it if necessary. The microcontroller is programmed using the user's pins, and the sensor data is then obtained.

4.7 Transformer health monitoring from traditional till modern techniques

Transformers are used mainly in the transmission and distribution of energy. The duty of dispersing the transformer under recharging circumstances (as specified on its nameplate) ensures a long service life. However, when they are subjected to severe loads, heating, low or high voltage/current, or unexpected failures, their life is significantly decreased. Excessive and insufficient transformer cooling is the primary source of dissatisfaction. Because the transformer is made up of many interconnected components, all of them should be inspected on a regular basis to keep the transformer in good working order.

As a result of the detailed analysis done over various areas as power distribution forecast and management of physical components of power distribution it is evidential that there is no compounded package available as a solution toward solving End—End problems occurs in various stages of power transmission till distribution. Since we have faced the occurrence of Day—Zero of coal and also it is feared that the war for power will happen in the near future we have proposed a unique solution which will serve its worth.

4.7.1 Transformer and distribution line health monitoring

AUnique sensor like fluid level sensor and temperature sensor are specially positioned inside the transformer's oil chamber to monitor and maintain defined oil temperature and oil level, which is the essential criteria in the deciding transformer health. Based on the temperature level and oil level the IoT and AI combo will activate the automated transformer cooling and oil cooling system. In addition, it will alert the engineer about the criticality, if the unwanted satiation raised (Fig. 4.1).

4.7.2 Power quality and test analysis

This system also acts as a model in assuring power quality as well as a detection and monitoring system for round the clock power theft watch

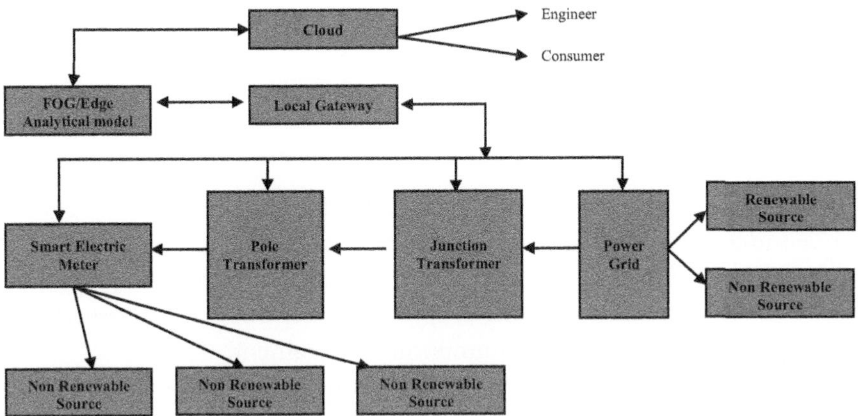

Figure 4.1 IoT based smart architecture for power distribution. *IoT*, Internet of things.

dog timer. It will assess the power transmitted through the TL in terms of its voltage and ampere, if there is any fluctuation in terms of meter reading then the system will check the health of the TL for detecting any blackouts brownouts and continuity of the is also test. In ordered to verify any disconnection in the physical line. If all the above set scenario are found negative then the system will audit the previous junction voltage and ampere reading to verify there is any line tapping has happened [10]. This reverse testing mechanism will help in locating the tapped junction and also intimates the junctional engineering about the occurrence of the power theft. So the process is done by AI based model which is debited specially for this cause in the Fog layer and the communication with nodal engineer will happen through the cloud and the cloud also acts as a storage medium which stores the copy data for future reference.

4.7.3 Demand and supply chain

The demand and supply management model is explicit positioned in the Fog layer to analysis the demand of the consumer through the deep learning prediction based algorithm and current consumption is uploaded in the cloud for future reference and also will be enhanced as a historical data. This system will also audit whether the supply is to the par of demand. Here, smart meters are used to identify the discriminate consumption of electricity and the data from the smart meter transmitted to the cloud through the aggregator.

4.7.4 Generation and distribution management of power

In this proposed model, the aim is to not only monitor the supply of the electricity but also monitoring the physical health as well as consumption pattern, pricing policy and also meeting the demand through generation or determining feasible alternate solution [11].

With the advent of integrating IoT-Fog, EDGE, AI, and ML together in defining optimistic feasible and cost-effective end—end problems occur in power generation, distribution, and consumption. The following are the possible outcomes that this proposed model offers to society toward quality quantitative power consumption management.

1. This system proposes a dual band consumption-based pricing policy (peak and non-peak hour pricing policy), which will encourage the consumer toward prioritize they're during the peak hours, which will reduce peak hour demand to a greater extent.
2. This system also performs a discriminative analysis on individual homes which suggest the consumers toward revamp their usage. So, this dual band pricing policy encourage the consumer through awards and rewards. The proposed model handful in generation and distribution of power as per the demand of the consumers. It also stock analysis a well as aiding support and power grid management and also helpful in switching to renewable resources such as solar, wind, etc. It also ensures the quality of power been supplied to the consumers.
3. This system is smart enough in prediction of false as well as locating the area of occurrence. In overall perspective this system well a compounded package for the problems occurs because of power.

4.7.5 Live time on a day

The live time on a day (LTD) tariff [12] is widely recognized in the electricity industry as an important demand supply management (DSM) criterion for exponentially growing customers to shift a fraction of their massive amounts from peak to off times, attempting to improve the system load factor by lowering demand on the system during peak hours. Consumption tariff chart [12] is summarized in Table 4.1.

LTD tariffs give clients economic signals that reflect the complete cost of manufacturing, delivering, and distributing power, allowing resources to be effectively allocated. More price-based demand response, especially during peak power grid hours, could assist reduce or adjust consumer demand [13]. As a result, LTD tariffs are important in terms of spreading

Table 4.1 Consumption tariff chart [12].

Unit consumption	Non-peak hour pricing (in Rupees per unit/per hour)	Peak hour pricing (in Rupees per unit/per hour)
Less than 20 units	1	Not applicable
20–50 units	1	NPHP + 1.5
50–100 units	1	NPHP + 3.5
100–150 units	1	NPHP + 5.5
Greater than 150 units	1	NPHP + 7.5

Table 4.2 Consumer reward chart [14].

Consumer ranking	Award for first quarter	Award for second quarter	Award for third quarter	Award for fourth quarter	Billing year
1–100	100% waiver	100% waiver	100% waiver	100% waiver	1 g of gold coin
100–200	50% waiver	50% waiver	50% waiver	50% waiver	25 g of silver coin
200–300	25% waiver	25% waiver	25% waiver	25% waiver	Travel bag

and implementing DSM as well as achieving energy efficiency in the country. Peek consumption pricing and non-peek consumption pricing has been proposed to educate and create interest to the consumers toward shifting their non-important or toward switching to the locally available renewable power resources.

So based on the billing, the consumers who are consuming less during the peak hour are rewarded with gifts, goodies or waiver of total bill in every quarter of a year, the following table illustrates a pricing policy toward the production and consumption. This kind of pricing policy will always entertain consumers toward energy saving during peak demand hours that will directly reduce the requirement of whole power. Table 4.2 summarizes the consumer reward chart [14].

By raising tariffs [14,15] at peak hours and lowering them during off-peak hours, dynamic rates aim to reduce utility system expenses and optimize consumer bills. Load shaping aims to reduce peak loads and/or shift them to off-peak hours. Furthermore, if a consumer is unable to respond to a pricing signal, the pricing signal's original purpose may be overlooked. This is the most common circumstance where a client's usage is very low and the company has little flexibility to transfer or lower the load, and the customer has a constant daily energy consumption. Aside from being a DSM policy, the LTD tariff also has the purpose of assisting in the establishment of pricing structures for emerging technologies of generation plants that can encounter system peaking power requirements, which would be especially important in light of our country's peak demand deficit scenario.

4.8 Conclusion

To conclude, for the demand forecast analysis, there are various forecast algorithms such as statistics-based algorithm till RNN-based algorithms are tried at various stages. Similarly, semi-automated power consumption and distribution monitoring of power supply are done using low level processors up to modern on chip systems. Still, there is no proper system in the consumer side power supply management and no stranded power distribution policy that enhances the supply of power throughout the nation as same. It is found conclusively that there is no dissipated infrastructure for consumer side physical health quality and theft monitoring. The proposed model with the advert technologies such as IoT, AI, and ML can assure a complete solution of power management at various levels of distribution from generation (G) and transmission (T) till utilization (U).

References

[1] https://en.wikipedia.org/wiki/Electric_power
[2] D. Srivastava, M.M. Tripathi, Transformer health monitoring system using Internet of things, 2018 Second IEEE International Conference on Power Electronics, Intelligent Control and Energy Systems (ICPEICES), IEEE, October 22, 2018, pp. 903–908.
[3] L.K. Narayanan, P. Subbaiyah, I. Gururajan, R. Sampathkumar, R.R. Muralidharan, IoT-fog integrated voice of the plant based smart irrigation and power management system for smart farming, NVEO-Natural Volatiles & Essential Oils Journal| NVEO (2021) 5411–5421.
[4] H. Malik, R. Sharma, EMD and ANN based intelligent fault diagnosis model for transmission line, Journal of Intelligent & Fuzzy Systems 32 (4) (2017) 3043–3050.
[5] A. Jeevanandham, B. Nitin, M. Maheshkumar, B. Monish, Solar powered real time transformer health monitoring system using Internet of things (IoT), Journal of Physics: Conference Series 1916 (1) (2021) 012155. IOP Publishing.
[6] K.M. Ghori, R.A. Abbasi, M. Awais, M. Imran, A. Ullah, L. Szathmary, Performance analysis of different types of machine learning classifiers for non-technical loss detection, IEEE Access 8 (2019) 16033–16048.

[7] S.R. Fahim, S.K. Sarker, S.M. Muyeen, S.K. Das, I. Kamwa, A deep learning based intelligent approach in detection and classification of transmission line faults, International Journal of Electrical Power & Energy Systems 133 (2021) 107102.

[8] N. Shabbir, R. Ahmadi Ahangar, L. Kütt, M.N. Iqbal, A. Rosin, Forecasting short term wind energy generation using machine learning, 2019 IEEE 60th International Scientific Conference on Power and Electrical Engineering of Riga Technical University (RTUCON), IEEE, October 7, 2019, pp. 1−4.

[9] J.R. Wijesingha, B.R. Hasanthi, I.P. Wijegunasinghe, M.K. Perera, K.T. Hemapala, Smart residential energy management system (REMS) using machine learning, 2021 International Conference on Computational Intelligence and Knowledge Economy (ICCIKE), IEEE, March 17, 2021, pp. 90−95.

[10] P. Vadda, S.M. Seelam, Smart Metering for Smart Electricity Consumption [Internet] (Dissertation), 2013. Available from: http://urn.kb.se/resolve?urn = urn:nbn:se:bth-2476.

[11] P. Razmi, M.O. Buygi, M. Esmalifalak, A machine learning approach for collusion detection in electricity markets based on Nash equilibrium theory, Journal of Modern Power Systems and Clean Energy 9 (1) (2020) 170−180.

[12] Assignment on Implementation & Impact Analysis of Time of Day (TOD) tariff in India, Pricewaterhouse Coopers India Private Limited Regulatory Economics Advisory.

[13] A. Golder, J. Jneid, J. Zhao, F. Bouffard, Machine learning-based demand and PV power forecasts, 2019 IEEE Electrical Power and Energy Conference (EPEC), IEEE, October 16, 2019, pp. 1−6.

[14] N. Javaid, A. Ahmed, S. Iqbal, M. Ashraf, Day ahead real time pricing and critical peak pricing based power scheduling for smart homes with different duty cycles, Energies 11 (6) (2018) 1464.

[15] L. Dias, M. Ribeiro, A. Leitão, L. Guimarães, L. Carvalho, M.A. Matos, et al., An unsupervised approach for fault diagnosis of power transformers, Quality and Reliability Engineering International 37 (2021) 0748−8017.

CHAPTER 5

A survey on AI- and ML-based demand forecast analysis of power using IoT-based SCADA

Lakshmi Kanthan Narayanan[1], Priyanga Subbiah[4],
R. Rengaraj Alias Muralidharan[3], Arun Prasad Baskaran[2],
Venkatasubramanian Srinivasan[2], Aravind Prasad Baskaran[2],
Punitha Victor[2] and Hema Ramachandran[2]

[1]Department of Computer Science and Engineering, Quannta School of Software Engineering, Villupuram, Tamil Nadu, India
[2]Department of Computer Science and Engineering, Saranathan College of Engineering, Trichy, Tamil Nadu, India
[3]Department of Information Technology, Saranathan College of Engineering, Trichy, Tamil Nadu, India
[4]Department of Networking and Communication, SRMIST, Chennai, Tamil Nadu, India

5.1 Introduction

Since we are living in the technological era of Industry 4.0 and marching toward Industry 5.0, it, therefore, depicts that there is a huge demand of electric power at present and in the near future. In the recent past, we were aware of the power crisis evoked in the Republic of China, and it is feared that the shortage of power may get impended.

It is important to utilize the precious resource called energy in an appropriate manner, so that it paves way for our younger generation to get inspired as well. As we failed to identify the alternate energy generation resource in advance, indefinite closure of industry has occurred as a result of the aforementioned coal crisis. Drastic increases in population size is one major reason for the increase in unaccountable demand hike and on the other hand, the usage of electrical gadgets in our day-to-day life become inevitable [1]. Electric vehicles have the potential to reshape the transportation sector; whose insistence directly alarmed the human community about the demand of electricity is creeping up in an unimaginable spike. Therefore, the need for power audit is highly essential to combat the shortcoming power scarcity. The day zero of power has already raised the alarm for the emergency in the immediate damage control needed to be made toward power audit and supply [2].

Smart Energy and Electric Power Systems
DOI: https://doi.org/10.1016/B978-0-323-91664-6.00010-3

In the modern electricity distribution system, there is no proper defining method to analyze the consumer demands, and the distribution is not carried out in an optimal manner. The traditional way of performing the demand forecast analysis has implemented statistical forecast methods such as Time Series Algorithms (TSA), followed by Regression-based algorithms. Though these statistical models are used in performing the demand forecast of electricity, yet the accuracy of prediction has estimated to be around 50%–60%, leading to a lapse of 40% shortage in the efficiency of forecast lags [3].

This chapter is organized as follows. Section 5.2 discusses about the various Supervisory Control and Data Acquisition (SCADA)-based power consumption monitoring and metering systems in detail. Section 5.3 interprets the traditional power demand prediction method. Section 5.4 elaborates a machine learning (ML)-based demand forecast analytic model. Section 5.5 discusses the deep learning-based demand forecast analytic model in details. Section 5.6 briefs out overall inferences derived from the study of various forecast techniques. Section 5.7 concludes with the efficient power management policy adoption.

5.2 SCADA-based power consumption monitoring and metering systems

Programmable Logic Controllor (PLCs) are used to continuously monitor the distribution components and communicate the data to a centralized PC-based SCADA system. This technique can be used to find out the location of a fault in the event of a pipe break or damage. The SCADA controls [4] the feeder and distribution valves in the power distribution network. SCADA also provides the operators with alerts and can shut down the system if a critical scenario arises [5]. The electrical signal character of power's conductivity has been employed as a quality distribution management metric, SCADA systems are divided into generations, which are explained below.

5.2.1 First-generation SCADA system

Through the ages when the computer networks were not in existence, the SCADA systems were developed with monolithic control architecture [5]. These SCADA systems are standalone system, which are not interconnected to any other systems. The computational power of these systems is very low. Fig. 5.1 illustrates the first generation of SCADA.

5.2.2 Second-generation SCADA system

Unlike the first-generation SCADA, the cost and time involved in this model is combated to a certain extent [6]. As the network protocols are proprietary in nature, the communication network protocols are not standardized. Fig. 5.2 demonstrates the second generation of SCADA.

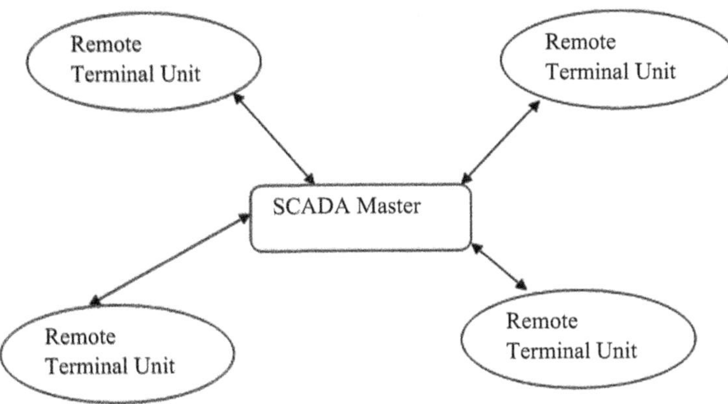

Figure 5.1 First-generation SCADA system. *SCADA*, Supervisory Control and Data Acquisition.

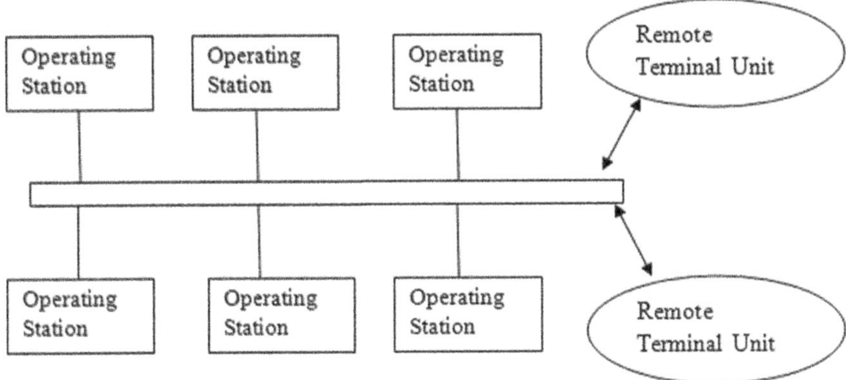

Figure 5.2 SCADA system of the second generation. *SCADA*, Supervisory Control and Data Acquisition.

5.2.3 Third-generation SCADA system

SCADA of the third generation is framed with simpler components in which the standardized communication protocols are introduced. The cost effectiveness of the large-scale system is brought through interconnection of larger geographical interconnection. Fig. 5.3 depicts the third generation of SCADA.

5.2.4 Fourth-generation SCADA system

The improvement in the SCADA communication technology results in the improvement in processing efficiency. The Internet of Things (IoT) systems can use open network protocols. Fig. 5.4 illustrates the fourth generation of SCADA.

5.3 Traditional power demand prediction method

The demand side response (DSR) values [8] over a course of time are anticipated and given in Table 5.1 as follows. The observation made on the upper bound and lower bound responses of demand side is illustrated in the following Fig. 5.5.

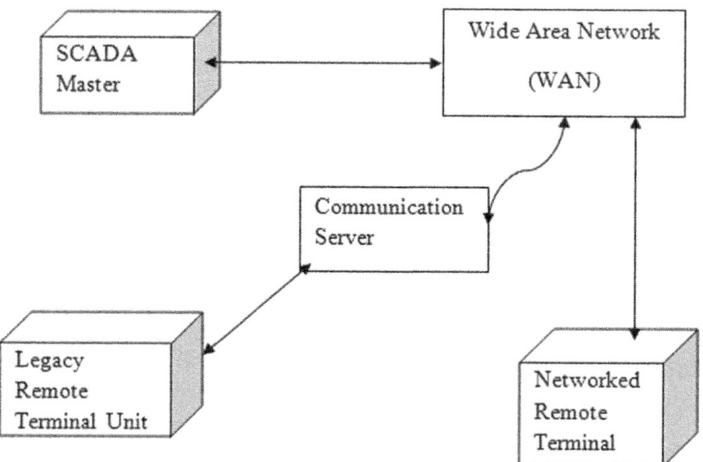

Figure 5.3 Third-generation SCADA system. *SCADA*, Supervisory Control and Data Acquisition.

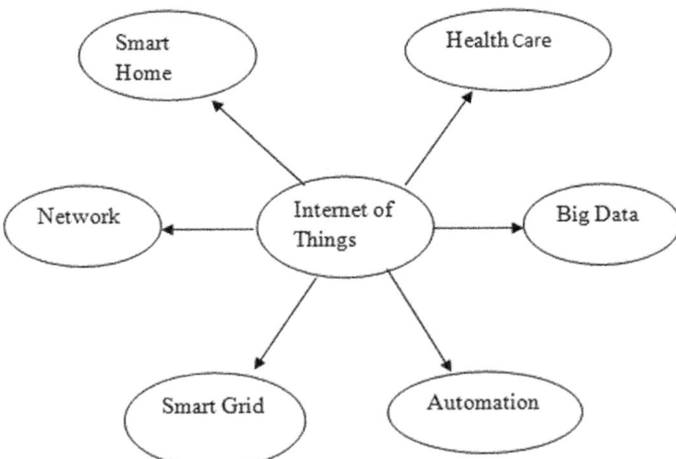

Figure 5.4 Fourth-generation SCADA system [7]. *SCADA*, Supervisory Control and Data Acquisition.

Table 5.1 Future DSR's [8].

Time	Upper bound of DSRs	Lower bound of DSRs
20170101	5212.24	5027.60
20170102	5786.46	5579.83
20170103	6440.54	6248.74
20170104	6828.80	6543.81
20170105	7143.77	6936.57
20170106	7257.46	7045.09
20170107	7289.65	7139.05
20170108	7255.56	7112.50
20170109	7298.15	7086.67
20170110	6953.14	6805.47

DSR, Demand side response.

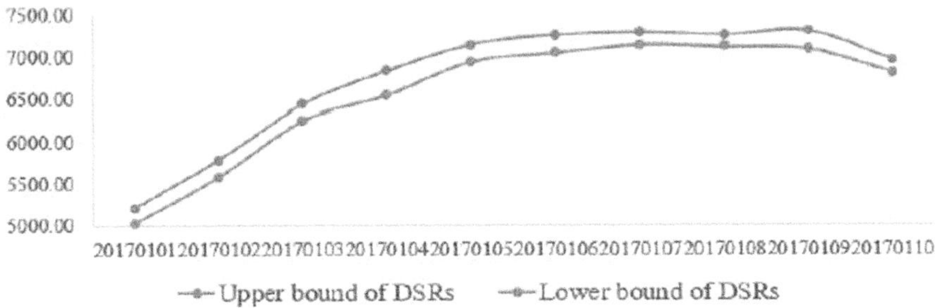

Figure 5.5 Limits of DSR [9]. *DSR*, Demand side response.

5.3.1 Time series analysis using ARIMA and S-ARIMA (DSR)

The first step in forecasting Error $e_t = Y_t - F_t$ is used to calculate the error [10]. There are n error terms for "n" periods, where the statistical measures are given by:

$$\text{Mean Error ME} = \frac{1}{n}\sum_{t=1}^{n} e_t \tag{5.1}$$

$$\text{Mean Absolute Error MAE} = \frac{1}{n}\sum_{t=1}^{n} |e_t| \tag{5.2}$$

$$\text{Mean Square Error MSE} = \frac{1}{n}\sum_{t=1}^{n} e_t^2 \tag{5.3}$$

$$\text{Percentage Error PE} = \left(\frac{Y_t - F_t}{Y_t}\right) \times 100 \tag{5.4}$$

The relative measures are used frequently

$$\text{Mean Percentage Error MPE} = \frac{1}{n}\sum_{t=1}^{n} \text{PE}_t \tag{5.5}$$

$$\text{Mean Absolute Percentage Error MAPE} = \frac{1}{n}\sum_{t=1}^{n} |\text{PE}_t| \tag{5.6}$$

In this ARIMA and S-ARIMA, the forecast accuracy was obtained within a range of 50%−60%. This is not a big number and definitely needs improvement in the forecast accuracy [11,12]. So, the next level of demand analytic model was implemented on bases of regression. Prediction accuracy toward electric power demand assessment is around 45%−50%.

5.3.2 Linear regression

For the comparison forecast analysis [13], a ML–based regression model is considered. Least Square Linear Regression is the model that was utilized (LSLR) which is used to calculate the dependencies and independencies of two or more variables [14,15]. The month's date is one variable in this data collection, and the amount of consumption on that particular day is another.

The relationship between the variables x and y is represented by the slope of a straight line [15]. The straight-line regression computational model is represented by Eq. (5.7).

$$s = p + rX \qquad (5.7)$$

where "p" is referred as "s" intercept, Slope is represented as "r"

The above graph Figs. 5.1, 5.2, and 5.5 illustrates the deviation between forecast analysis made for the city of Austin, USA over a period of year in kilowatt hour. Here, the regression value should between 0.1 and 1.0. So, that the prediction model will be considered as a best fit of regression [16]. The symbolic way of representing the best fit of regression is represented as 2R.

$$R^2 = 0.1 \geq 1.0 \qquad (5.8)$$

By deploying regression–based model toward demand forecasting, it is inferred that accuracy is proved to be around 60%–70%. Again, this efficiency is not enough for such crucial analytics [17]. To increase the forecast model's performance, it was hypothesized and implemented with the evolution of ML-based prediction.

5.4 Machine learning-based demand forecast analyst model

Renewable Energy Sources are the main target of recent analysis because of the chances they supply to the facility system and the challenges they create. Wind energy generation prediction could be an important facet of installation and also the difficulties of leveling provide and demand [18]. In this chapter, a regression rule supported support vector machine (SVM) to predict the wind energy production in Estonia in some unspecified time in the future ahead. Despite of the most important edges of alternative energy, foregone conclusion could be a tough task. Wind power generation is stricken by factors besides the native season and climate, and also in the presence of chaotic turbulence makes prediction harder. The ML algorithms are important tools for alternative energy generation prediction [19]. They were used for short-run prediction and may be extended for long prediction once enough knowledge is obtainable. In 2018, the common share of renewable energy in Estonia was 29 p.c. it's nearly 100% on top of the common of Europe countries' share of renewable energy, and it's on top of the EU's target price for 2020.

The energy forecast is then calculated through victimizing this model by scrutinizing the actual prediction, the prediction created by Elering's rule, and also the prognostication created by the projected SVM-based model. The results clearly showed that the SVM-based model produces far better results that par on the point of the particular energy generation values. When compared to Elering's rule, the projected rule includes a 100% lower RMSE price. The prediction results are solely valid for in some unspecified time in the future. This model will be extended to predict weeks and months ahead within the future; a neural network-based prediction rule may be developed and compared to the current one.

Collusion detection criteria square measure computed for the equilibrium and their peripheral points. Simulation results show that the accuracy of the used machines is suitable in collusion detection. The impact of every criterion on the collusion detection accuracy is evaluated and also the rate and Celtic deity indices play additional vital roles in collusion classification performance than different criteria. The results show the collusion of two or more additional corporations for a protracted period of time, the collusion is valid [20]. However, if the results show the collusion happens solely in associate degree hour, it can't be thought about as collusion.

Electricity consumption is increasing every day in every corner of the globe, resulting in a supply-demand imbalance. When we look at the world's electricity generation by energy sources, we find that fossil fuels account for 63.3%. In recent decades, energy management has played a key role in both the sectors at the utility side (supply side) and the household level (demand side). Demand Facet Management (DSM) is a critical load management action at the customer end [10,21]. Load management encompasses load shifting, load programming, and other demand-side energy management approaches, but it's also linked to other ML techniques at the moment.

In recent years, number of scholars focused on good REMS in a variety of methods, coming up with a variety of scientific techniques to design and develop a sophisticated energy management system combined with an extremely good grid framework. It responds to a total power outage in a district by leveling the provision facet with demand-side stated, and the planned system substitutes the partial load shedding theme in an extremely manageable manner. Furthermore, the system's choice-making process takes the buyer's load preference into account. The analysis gap with our proposed system is uninterruptible power to all or any hundreds at the largest extent possible, the use of renewable greener approaches with ML, and the use of Wi-Fi connection to cover long distances. Using ML techniques, the proposed system (REMS) in this work effectively switches hundreds between the most grid and renewably energized native battery storage energized by top star at the household premises. Owing to a few difficulties, the battery is charged entirely by the star PV. The battery, in particular, does not receive the greatest amount of grid electricity.

The classifiers are still performing admirably. ML approaches are also highly beneficial in determining to spot real offenders, unlike other network-based systems, which are simply capable of identifying a section where NTL is committed but fail to discover the stolen incidents. Furthermore, the value of abuse ML methodologies is significantly lower when compared to other techniques like as manual on-the-spot inspection. For NTL recognition, several ML algorithms are being tried, including recently found ensemble techniques such as CatBoost, LGBoost, and XGBoost.

By the advert of implementing ML-based forecast analysis done on consumer demand prediction, the performing efficiency of the ML based is arrived close to an average of 85%—90%. Still there is some more improvement is possible in terms of prediction, toward clocking the prediction accuracy to the maximum using deep neural-based algorithms.

5.5 Deep learning-based demand forecast analytic model

Deep learning-based algorithms are the result of RNN, in which the output of the present stage will be fed as a partial input to next stage.

The efficient operation of a power system, both in power systems and in everyday life, depends on its safety and dependability. Transient stability is a critical obstacle that, if properly examined, can aid in the continued operation of this secure and trustworthy system. Data mining and ML tools have recently become popular for assessing Transient Stability (TSA). During the operation of a power system [22] disruptions such as cut-off loads, short-circuit faults, and so on may occur, and it is vital to be able to rapidly and accurately evaluate if the power system is stable following these disturbances. TSA's purpose is to meet the demands for speed and capacity, as well as calculating accuracy and user-friendliness.

Transient Stability refers to a power system's capacity to retain synchronism after suffering interruptions. During power system operation, real-time TSA is utilized to allow for corrective actions to be taken before, during, and after a fault. When a transient occurs, this enables tight prediction of stability, allowing preventative or corrective modifications to be implemented as soon as possible. The stability of the 19-bus power system is projected using DL based on long short-term memory (LSTM).

Intelligent power mechanisms have been proposed to provide the best solution for the coordination, balance, and control of the entire energy system. A novel approach of energy equation and communication in the user side electricity sector, peer-to-peer (P2P) electricity transfers, can effectively promote energy sharing within the user group and boost the financial benefits of users participating in the energization process [23]. Reinforcement learning (RL) is an artificial intelligence (AI) method in which agents constantly obtains applicable experience and expertise whereas conversing with their surroundings, automatically adjust their decision-making actions, and achieve maximum their return.

A peer-to-peer electricity transaction is established for low transaction participants. The supply and demand ratio (SDR) is used to evaluate the transaction price based on basic economic ideas. It can be calculated and updated in real time without the involvement of users. It is better suited to the characteristics of small-scale peer-to-peer markets. The electricity payment problem is similar to an MDP, and RL elements such as state space, action space, reward function, and transition rate are investigated. In MDP, the ESS actions and transaction strategy selection problems are average quietly processed, and the SARSA online RL algorithm is used to analyze and solve them. It is possible to realize the RL solution of the ESS action selection and transaction decision problem for small-scale P2P electricity transaction participants.

5.6 Summary

Figs. 5.6, 5.7, and 5.8 illustrates how the integration of IoT, Fog, and ML as a consumption data collection and demand forecast analytic model, which can be a complete solution for the current power crisis, which will educate the consumer about their consumption and also the supplier board, is furnished with the futuristic demand so that they can make precautionary requirement [24]. The meted consumption from the various

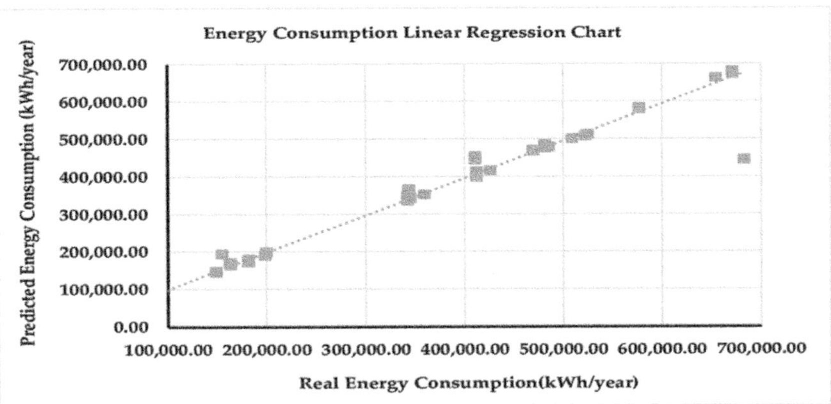

Figure 5.6 Energy generation SCADA systems [8]. *SCADA*, Supervisory Control and Data Acquisition.

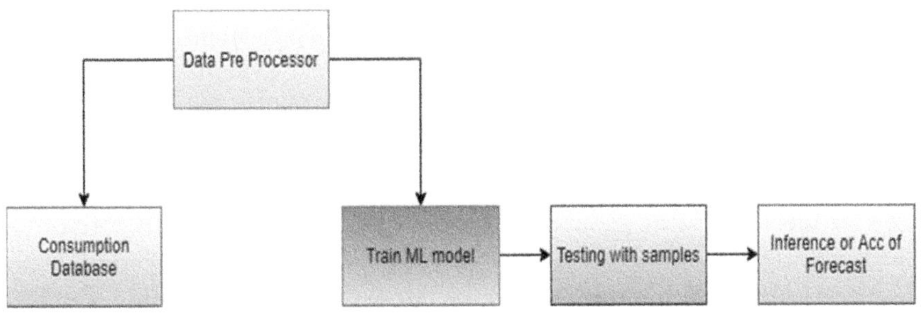

Figure 5.7 IoT—Fog-based electricity forecast model.

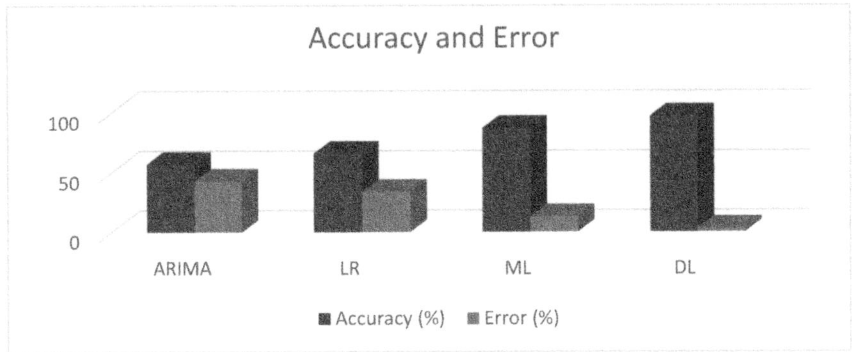

Figure 5.8 Comparative analysis of forecast efficiency.

houses is collected by aggregated and forwarded to the edge for prediction analysis. Forecasted data are updated in the cloud, and it will be forwarded to the electric board, which will help in proceeding with the necessary possible action toward serving as per the demand (Figs. 5.7 and 5.8).

The average prediction accuracy of various consumption forecast algorithms is summarized in the graph below. We may conclude that the recurrent neural network (RNN)-based deep learning algorithm has the best prediction accuracy based on the findings of the inquiry.

5.7 Conclusion

On the detailed survey of various demand forecast analytic technique from the traditional statistical-based model till the modern deep learning analytical model, it is inferred that the modern forecast model performs far better than that of the traditional statistical and regression-based models. The forecast accuracy is inferred more than 95% by the ML and deep learning forecasting algorithms, but still the modern techniques are black boxed models, and the process can't be structured evidently. At the same time, the traditional models are white boxed model in nature, where the forecast process is defined. It is a need of the hour to implement AI- and ML-based demand forecast model, to satisfy the consumer for power. This in terms helps in generating power or finding the alternate resource of power by the distribution company. It will be highly quantitative for predicting the consumer demand for a long short term period accounting hourly demand for power throughout the days. This will help in building an IoT−Fog-based power distribution and management system in framing a smart grid. The prediction accuracy toward electric power demand assessment is around 90%−97%.

References

[1] J. Bo, S. Chen, L. Qiao, S. Liu, K. Tian, H. Wang, et al., Extraction and analysis of heat storage load characteristics for power quality improvement, 2019 IEEE International Conference on Signal, Information and Data Processing (ICSIDP), IEEE, December 11, 2019, pp. 1−5.
[2] J. Cui, T. Qi, Y. Zhu, K. Xie, X. Ma, Modeling and control of multiple flexible loads in demand response considering renewable energy accommodation 2020, IEEE 3rd Student Conference on Electrical Machines and Systems (SCEMS) IEEE December 4 (2020) 908−912.
[3] K.M. Ghori, R.A. Abbasi, M. Awais, M. Imran, A. Ullah, L. Szathmary, Performance analysis of different types of machine learning classifiers for non-technical loss detection, IEEE Access 8 (2019) 16033−16048.

[4] Y. Zhang, L. Wang, Y. Xiang, C.W. Ten, Power system reliability evaluation with SCADA cybersecurity considerations, IEEE Transactions on Smart Grid 6 (4) (2015) 1707−1721.

[5] H.M. Hosseinzadehtaher, A. Khan, M.B. Shadmand, H. Abu-Rub, Anomaly detection in distribution power system based on a condition monitoring vector and ultra-short demand forecasting, 2020 IEEE CyberPELS (CyberPELS), IEEE, October 13, 2020, pp. 1−6.

[6] Y. Hu, Y. Sun, C. Yu, Z. Pan, X. Yuan, A load forecasting algorithm for the demand side resources, 2021 IEEE International Conference on Power Electronics, Computer Applications (ICPECA), IEEE, January 22, 2021, pp. 418−422.

[7] https://stid.com/en/technology/internet-of-things-iot.

[8] T. Singh, A.S. Sethi, H. Singh, P.S. Rana, N. Kumar, M.S. Obaidat, Two-tier ensemble model for demand side prediction in smart grid environment, 2019 IEEE Global Communications Conference (GLOBECOM), IEEE, December 9, 2019, pp. 1−6.

[9] R.J. Tom, S. Sankaranarayanan, J.J. Rodrigues, Smart energy management and demand reduction by consumers and utilities in an IoT-fog-based power distribution system, IEEE Internet of Things journal 6 (5) (2019) 7386−7394.

[10] S.Q. Bu, W. Du, H.F. Wang, Hypothesis testing of the stochastic model of demand and supply power of plug-in electric vehicles, International Conference on Sustainable Power Generation and Supply (SUPERGEN 2012), IET, September 8, 2012, pp. 1−6.

[11] Z. Shi, H. Chen, Z. Min, F. Bai, Diversified power supply model based on analytic hierarchy process, 2018 IEEE PES Asia-Pacific Power and Energy Engineering Conference (APPEEC), IEEE, 2018, pp. 39−44.

[12] M. Li, H. Liu, T. Luo, Y. Liu, X. Li, Deep reinforcement learning based reliability pricing strategy in electricity spot market, 2020 International Conferences on Internet of Things (iThings) and IEEE Green Computing and Communications (GreenCom) and IEEE Cyber, Physical and Social Computing (CPSCom) and IEEE Smart Data (SmartData) and IEEE Congress on Cybermatics (Cybermatics), IEEE, 2020, pp. 901−909.

[13] O.I. Adebisi, I.A. Adejumobi, Development of a load management scheme for the Nigerian deregulated electricity market using regression model, 2019 IEEE PES/IAS Power Africa, IEEE, 2019, pp. 682−687.

[14] K. Goswami, A.B. Kandali, Electricity demand prediction using data driven forecasting scheme: ARIMA and SARIMA for real-time load data of Assam, 2020 International Conference on Computational Performance Evaluation (ComPE), IEEE, 2020, pp. 570−574.

[15] H. Wooten, S. Harris, B. Ramachandran, An Innovative Deep Learning Approach Applied to Transient Stability Assessment of Power Systems, 2020 Clemson University Power Systems Conference (PSC), IEEE, 2020, pp. 1−6.

[16] S. Kanwal, B. Khan, M.Q. Rauf, Internal power reserve estimation for frequency regulation in storageless photovoltaic system, 2019 22nd International Multitopic Conference (INMIC), IEEE, 2019, pp. 1−6.

[17] Z. Mubarok, M. Langie, S. Soeyati, Electricity demand forecasting using a simple-e expanded approach at PT PLN (Persero) of kotamobagu area from 2018 to 2022, 2019 International Conference on Technologies and Policies in Electric Power & Energy, IEEE, 2019, pp. 1−6.

[18] A. Golder, J. Jneid, J. Zhao, F. Bouffard, Machine learning-based demand and PV power forecasts, 2019 IEEE Electrical Power and Energy Conference (EPEC), IEEE, 2019, pp. 1−6.

[19] P. Razmi, M.O. Buygi, M. Esmalifalak, A machine learning approach for collusion detection in electricity markets based on Nash equilibrium theory, Journal of Modern Power Systems and Clean Energy. 9 (1) (2020) 170—180.

[20] N. Shabbir, R. AhmadiAhangar, L. Kütt, M.N. Iqbal, A. Rosin, Forecasting short term wind energy generation using machine learning, 2019 IEEE 60th International Scientific Conference on Power and Electrical Engineering of Riga Technical University (RTUCON), IEEE, 2019, pp. 1—4.

[21] J.R. Wijesingha, B.R. Hasanthi, I.P. Wijegunasinghe, M.K. Perera, K.T. Hemapala, Smart residential energy management system (REMS) using machine learning, 2021 International Conference on Computational Intelligence and Knowledge Economy (ICCIKE), IEEE, March 17, 2021, pp. 90—95.

[22] Y. Ma, Q. Zhang, J. Ding, Q. Wang, J. Ma, Short term load forecasting based on iForest-LSTM, 2019 14th IEEE Conference on Industrial Electronics and Applications (ICIEA), IEEE, June 19, 2019, pp. 2278—2282.

[23] D. Wang, B. Liu, H. Jia, Z. Zhang, J. Chen, D. Huang, Peer-to-peer electricity transaction decision of user-side smart energy system based on SARSA reinforcement learning method, CSEE Journal of Power and Energy Systems (2020).

[24] L.K. Narayanan, P. Subbaiyah, I. Gururajan, R. Sampathkumar, R.R. Muralidharan, IoT-Fog integrated voice of the plant based smart irrigation and power management system for smart farming, NVEO-Natural Volatiles & Essential Oils Journal| NVEO. (2021) 5411—5421.

Further reading

Alshibani A, Prediction of the Energy Consumption of School Buildings, Applied Sciences. 10 (17) (2020)5885. Available from: https://doi.org/10.3390/app10175885. In press.

CHAPTER 6

Impact of artificial intelligence techniques in distributed smart grid monitoring system

R. Senthil Kumar[1], S. Saravanan[1], P. Pandiyan[2] and Ramji Tiwari[3]
[1]Department of Electrical and Electronics Engineering, Sri Krishna College of Technology, Coimbatore, Tamil Nadu, India
[2]Department of Electrical and Electronics Engineering, KPR Institute of Engineering and Technology, Coimbatore, Tamil Nadu, India
[3]Department of Electrical and Electronics Engineering, Sri Krishna College of Engineering and Technology, Coimbatore, Tamil Nadu, India

6.1 Introduction

The growing global population necessitates the expansion of facilities, forcing energy service providers to increase their production. However, fossil fuels dominate global power generation, and they are the primary source of carbon dioxide (CO_2) emitters in the atmosphere. Governments worldwide are promoting renewable electric energy sources to reduced CO_2 emissions from the conventional power grid [1]. For instance, the United States of America contributed 72 GW of solar and wind energy generation between 2018 and 2021, encouraging government tax incentives and lowering capital costs [2]. Across the globe, similar efforts are being made to boost the utilization of renewable sources. In this field, a variety of studies are carried out, and market suggestions are constantly changing. Renewable energy sources (RES) can, in line with their international environmental objectives, provide an alternative source to fossil fuel dependency by creating green energy options to reduce hazardous gas emissions and reduce peak load demand. Future power systems (PSs) can benefit from integrating RES using smart grid (SG) technologies.

The concept of SG is to change the current electrical grid to an electronically controlled network from an electromechanically managed system. The United States Department of Energy published a report on the SG system [3]. SG systems comprise digital-based sensing and field equipment's that work together to manage multiple electrical processes, communication methodologies, information management and control technologies. Conventional

Smart Energy and Electric Power Systems
DOI: https://doi.org/10.1016/B978-0-323-91664-6.00005-X

grid planning and operation challenges have been altered into three significant tasks due to the progress of SG concept. There is (1) the ability to monitor real-time processes, transmit data back to operating centers, and frequently react automatically when procedures need to be adjusted; (2) devices and systems can exchange data with each other; and (3) the data generated by digital technologies throughout the grid is processed, analyzed, and operators are assisted in accessing and applying the information. SG problems include power grid stability, SG security, and load forecasting and fault detection [4]. These critical features enable the gathering of huge volumes of high-dimensional data as well as multi-type data concerning the process of the utility control grid. Alternatively, conventional modeling, control strategies, and optimization have numerous limits in processing large datasets. As a result, artificial intelligence (AI) approaches in the SG are becoming increasingly visible.

AI approaches rely on large volumes of data to construct intelligent machines suitable for executing tasks that would otherwise need human intellect. The subdivision of AI is machine learning (ML). The terms ML and AI are used interchangeably in some cases. In contrast to this, ML is only one method of achieving AI. Fuzzy logic (FL), neuronal networks, expert systems (ES) and robotics are more general approaches to developing AI systems. In general, AI approaches to speed up and improve the accuracy of decision-making. AI in SG applications can be well-defined as computers that replicate grid operators' cognitive processes to achieve self-healing capabilities. In other circumstances, grid operators may not be replaced by AI. Although AI system can be more precise, reliable, and extensive, there are still numerous problems associated with integrating AI approaches into the SG. The SG supports 2-distinct kinds of AI systems: Physical AI and Virtual AI. Informatics is a component of virtual AI systems that can assist utility grid operators in their work. AI systems in the SG can be classified again into two types: Artificial Generic Intelligence (AGI) and Artificial Precise Intelligence (API). An AGI is an artificial intellect system designed to acquire and evolve independently, similar to human beings. On the other hand, an API is an AI system designed for specific jobs with unique criteria and limits like an AI system that does load prediction using several datasets. In the future, the development of AGI systems may contribute to the development of truly intelligent grid systems.

Because of the growing complexity of the energy industry, sophisticated processes are required to manage the energy system and make decisions on time properly. For the most part, AI techniques such as genetic algorithms (GAs), artificial neural networks (ANN), multiple agent systems, and reinforcement learning are familiar. They have handled problems such as forecasting, classification, optimization, control techniques, and networking [5]. Because of a shortage of automatic controlled resources, various system functions are mostly performed manually or semiautomatically. However, incorporating AI into the grid system would result in new grid guidance and improvements. Optimizing controllable loads through an AI approach leads to cost savings. To maximize the controlled loads, a GA was proposed in [6] to manage isolated microgrids (MGs). With the rise of available data storage and computing power, AI approaches can deal the constraints of the conventional grid system powerfully and effectively. Moreover, various security problems have been created by the application of distributed SG computing methods. Cyberattacks and physical attacks are dangers that can cause privacy breaches, infrastructure failure, service denial, and disruption.

The rest of this chapter is structured in the following manner. The future energy system is illustrated in Section 6.2. The distributed SG monitoring system powered by AI approaches is given in Section 6.3. Section 6.4 discusses the Integration of RES. Section 6.5 addresses the application of AI to the integration of energy storage systems (ESS). Section 6.6 discusses the economic aspects of SGs and the deregulation of the market. In Section 6.7, the difficulties and issues associated with implementing an AI-based distributed SG monitoring system are discussed. Finally, the conclusion was made in Section 6.8.

6.2 Future energy system

The present, aging, and congested distribution centers cannot provide continuous, high quality electricity in a safe and effective manner [1]. The electrical grid is experiencing a tremendous transition in intelligent approaches as demand and generation rise [7]. The future SG is a safe, ascendable, and real-time bidirectional flow of electricity constantly. The large scale integrating of distributed energy resources (DERs) into primary grid has altered the PS establishment and maintenance worldwide over the last twenty years. Also, utility service providers should manage variable generation for DERs that lack advanced intercommunicational facilities. SG is a viable option for improving the conventional electrical grid network by rigorously integrating Information and Communication Technology (ICT), enabling higher integration of distributed elements [8].

The eight key areas for SG standards, in accordance with the National Institute of Standards and Technology (NIST), are the following:

1. User's energy efficiency and demand response: Targets various user (consumer) categories to engage people inefficient energy use by regulating as well as planning their utilization patterns.
2. WASA (Wide-Area Situational Awareness): It offers grid operators exact data at the perfect time to make optimal decisions.
3. Electric transport systems: The various electric vehicles (EVs), such as PEV, BEV, and PHEV, provide cost-effective fuel, conserve the environment, improve living standards, and promote economic progress [9].
4. Energy storage: This technology saves energy for future use, allowing users to save money on power. It enables access to the grid and manages it by supplying backup for fluctuating RES.
5. Advanced metering: It collects and analyses data from smart meters and offers consumers a cost-effective management system.
6. Network communications: It uses bidirectional communication networks to connect smart energy devices.
7. Cyber security: It prevents different cyber-attacks on data acquired from the SG through ICT.
8. Distribution network management: It increases grid stability while decreasing losses.

The distributed Internet of Things (IoT) components use bi-directional high-speed communication systems to interact, analyze, and manage their resources individually or in conjunction with other equipment in a distributed and separate way. IoT and smart meters may help enhance the system's overall effectiveness, from primary load control in a home to complicated grid voltage stability control [10]. These devices can communicate with one another and self-learn to take action on their own. Due to rising digitalization in the power network, decentralized intelligent methods have enhanced system performance and dependability, including monitoring, service, fault detection, and RES incorporation. However, the SG has become increasingly complicated and susceptible to cyber terrorism as the number of dispersed devices grows, supporting technologies that allow multi-directional interaction across systems and devices. In this chapter, various AI methods in different applications in the SG, such as RES integration and energy storage system integration, are discussed. Furthermore, the chapter addresses the economic elements of the SG as well as market liberalization.

6.3 Distributed smart grid monitoring system using artificial intelligence

The AI technology in SG creates a digital framework for accessing solid technological capabilities. Energy, PS automation, examination of power demand statistics, and fault detection are all part of AI SG initiatives. The objective of an intelligent grid is to use AI to eliminate manual operations, which results in improved performance, reliability, and cost savings [11]. The electricity generation, the transportation of energy, the translation of power, energy supplies, and electricity consumption are all components of an electrical network. The following sections describe the use of AI in SG applications as shown in Fig. 6.1.

6.3.1 Electricity trading and online load forecasting

Load demand forecasting (LDF) enables real time power production and load management, which is now a necessary element of normal grid processes. The LDF is grouped into three categories: short, moderate, and long-term estimations, with periods ranging from a few minutes to more than 1 year [12]. One of the most often utilized techniques for predicting energy demands has been the ANN platform. Online analytical approaches are required to solve the data integrity issues and deliver accurate and rapid predictions. The stream is an accessible collection of data that can be recognized only once inside a restricted memory space. Non-static sets of data generate information that generally flows at fast rates. Online power load forecasting is critical to decreasing energy instabilities. Market connections can assist in closing the energy gap that exists between a contracted supply and actual use. AI can immediately adapt to massive amounts of data in the energy marketplace to acquire AI insights. Reasonable predictions may often assist in balancing the electrical system and ensuring sustainability. AI may be used to stimulate and enhance the incorporation of sustainable energy, particularly in predictions. As the degree of precision improves, ML and ANNs have been utilized to enhance estimations.

6.3.2 Fault detection and protection in the power grid

Dynamic electric tools refer to apparatus that concentrates on power electronics topologies. Dynamic power plants are used in alternating current transportation, direct communication, power storage, electrical distribution networks, regional grids, etc. To guarantee equipment safety, the

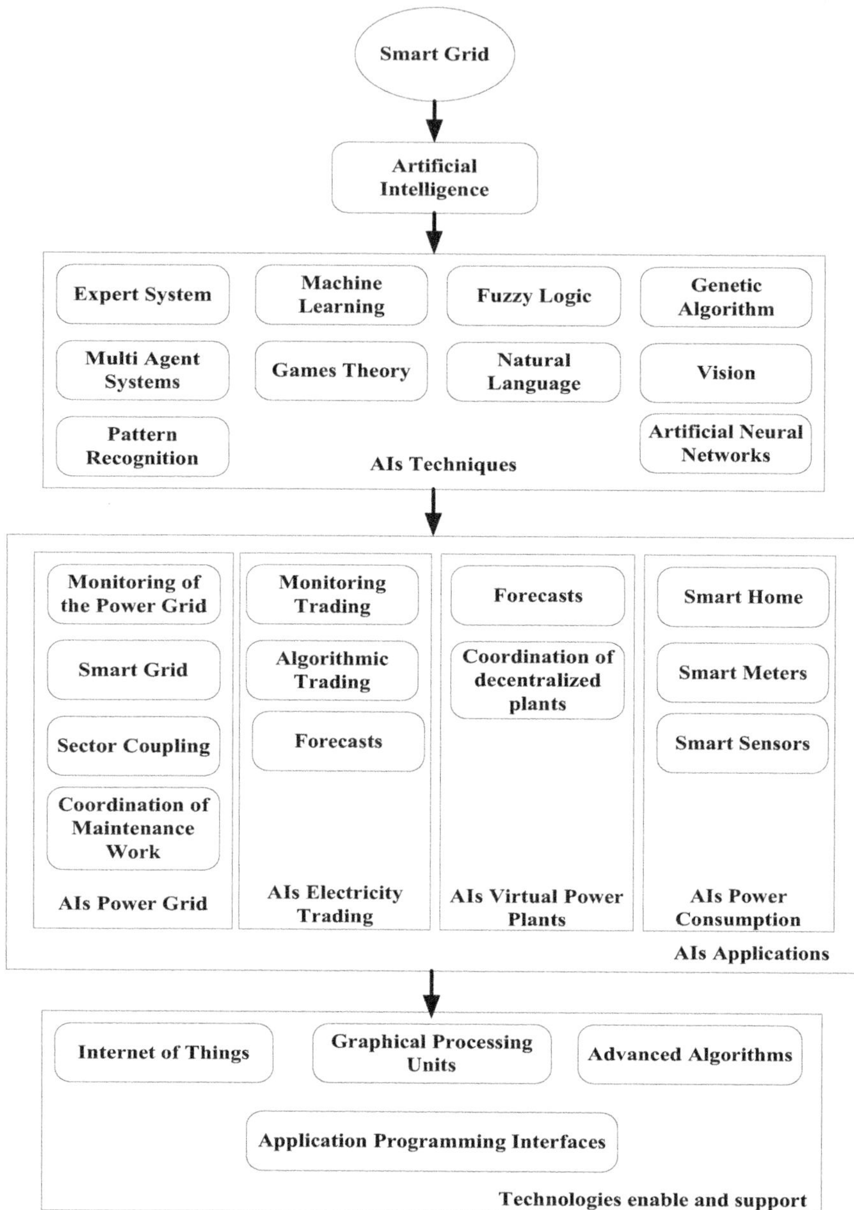

Figure 6.1 Artificial intelligence-based smart grid application.

condition monitoring and protection of robust PS equipment are essential [13]. The latter is critical in rapidly isolating problems, preventing machine degradation and the spread of defects. Due to its changing architecture, the tight coupling, unknown predictor variables, and other factors make it hard to identify the failures. AI with deep learning (DL) obtains in-depth characteristics for defect evaluations on flexible apparatus. It will offer additional information to enhance the data collection through learning immigration to transmit highly adaptable equipment's failure properties at many phases [14].

6.3.3 Consumer energy usage behavior

ML in AI can categorize non-uniform energy usage and forecast energy demand operations. The power consumption behaviors of distinct consumer groups can be identified using data collected by smart meters, AI assessment, and data mining. The statistics enhancement can be used to incorporate targeted services [15]. The precision of recognizing abnormalities is affected by environmental conditions, such as weather changes, variations in electrical equipment, and modifications in behaviors. Hence, it is essential to pay consideration to the negative features of the grid that could impact distribution across customers. Human characteristics are involved in energy consumption analysis, which can be solved through feature extraction or categorization. The development of DL multilayered hidden neural networks enhances the prediction performance of energy consumption and demand.

6.3.4 Power network protection in smart grid

Real time monitoring and adaptable power are part of SG network security. The detailed flow of information can assist the PS in resolving other difficulties. The power grid attack looks to be difficult to detect and shut down for a long period. DL technology can detect problems automatically, attack a network, and improve data security. The unusual data under attack on the targeted grid would be significantly fewer than the typical operational information. Therefore, it will be less than the training sample. Markers are not used in DL practice, which dilutes the quasi-sample sizes [16].

6.3.5 Distributed grid intelligence

Real time monitoring and adaptable power are part of SG network security. The detailed flow of information can assist the PS in resolving other difficulties. The power grid attack looks to be difficult to detect and shut down for a long period. DL technology can detect problems automatically, attack a network, and improve data security. The unusual data under attack on the targeted grid would be significantly fewer than the typical operational information. Therefore, it will be less than the training sample. Markers are not used in DL practice, which dilutes the quasi-sample sizes [16].

6.3.5.1 Distributed intelligence: prosumer side

A bidirectional electricity flow is attainable in SG due to advancements in the power infrastructure. Domestic energy consumers can generate and consume power as well as transfer it to other grid customers [17]. In the domestic, commercial, and industrial domains, millions of individuals exchange their RES. Fossil-fueled production will be replaced by an intelligent cooperative DER PS, in which consumers can trade electricity to maximize the financial benefits. AI techniques have emerged in the energy sector during the previous decade as a potential viable solution to provide optimum DERs and to assist reliable and fast demand response in real time. Energy providers aim for "all power grid choices to be made," from relay switches to massive generator controllers. To handle with unexpected and systematic disruptions, intelligent algorithms are therefore being created and implemented with forethought, self-learning and endurance.

AI researchers are still working on building operationally efficient algorithms that can properly forecast intelligent prosumer consumption and power data with real-time electrical pricing, which allows for beneficial electricity trading decisions. ES, FL, and ANN have used in the power industry to address technological difficulties, expense estimating, and defect identification in recent years owing to the fast progress of AI expertise. These approaches may also be used in domestic energy management (DEM), smart home demand-side management (DSM), and overall DSM [1].

6.3.5.2 Generation side distributed intelligence

The obstacles to today's modern power distribution networks are integrating dispersed energy sources, raising acceptance for RES adoption, developing proper planning, and establishing operational strategies that may grow demand while lowering emissions of greenhouse gases. The SG seems appears to be highly dependent on distributed system automated control, distributed energy storage capabilities, management, and PHEVs. This requires the development of more sophisticated and smart regulators, in addition to DERs, to supervise and operate the distribution grid. The monitoring and control of distributed generation have focused much on investigation and studies [1]. If a specific threshold is exceeded, the PS becomes unstable due to sustainability constraints. Energy management, failure identification and monitoring are all possible with distributed grid management.

The management of energy, power and convertors, and fault analysis are all part of the distributed functioning of PS designs. The SCADA system has traditionally been utilized to manage energy sources; however, this centralized design has proven to be almost unfeasible due to security as well as operational delays [18]. Since these systems generally demand human involvement for regular operations, they have grown less effective, significantly as the grid and its interconnections have evolved into more complex, requiring high speed and data management. In a distributed network, dispersed load balancing methods are meant to balance the various peers. As part of the load-balancing strategy, routers can communicate with each other and with DERs to move load to a low-consumption region from a high-consumption region. A more reliable and robust system is the result of this transfer. To regulate electrical networks, distributed state estimations and decentralized intelligent systems are used in conjunction with one another. In SG security, AI and blockchain technology can help with data storage and processing. Eck et al. [19] illustrate how AI methods are being deployed to help distribution grid operator's deal with extensive RES adoption through local energy trading.

6.4 Artificial intelligence techniques for the integration of renewable energy system

Energy supply and demand have grown rapidly as a result of global rural-to-urban migration. According to the United Nations, today, 56% of the global population resides in towns, that is expected to reach 68% by 2050. To meet the growing need for sustainable, clean, efficient, and secure electricity sources, it is necessary to integrate RES into the currently available PS network. In the coming years, the global proportion of RES in PSs could reach a spectacular ratio.

6.4.1 Consumer-side renewable energy sources integration

One of the primary elements of SG is the integration of RES and energy storage resources into consumer premises. Another essential feature is the sharing of the responsibility for managing energy flows and energy consumption [17]. In the case of RES, particularly solar, they generate energy on-site, reducing the need for long-distance transmission line losses as well as the need for huge capital investment and running costs.

Macedo et al. [20] proposed integrating photovoltaic and ESS with grid to reduce energy usage and improve efficiency. In conjunction with RES and ESS, SG systems efficiently regulate total energy consumption in high-energy-consuming buildings such as educational institutions, hotels, commercial buildings, and hospitals. For residential applications, Elkazaz et al. [21] have developed a hybrid fuel cell and solar PV with an optimization technique for the optimal real time process of DER.

The PS dependability is boosted by cost-saving and minimizing peaks for smart households with RES is reported in [22] when the home energy management system integrates cuckoo search algorithms, particle swarm optimization (PSO) and GA. Distributed energy is committed to maximizing RES and energy-producing technologies' utilization to maximize overall system efficiency and minimize environmental risks [23].

6.4.2 Generation-side renewable energy sources integration

As the temperature of the Earth continues to rise, the need for non-fossil fuel-based energy alternatives plays a significant role. RES is rising in popularity in the PSs sector because they are located closer to where they will be used. As a result of the increased deployment of RES, the system dynamics are transitioning to a new level, requiring storage facilities, two-way electricity flow, variable energy supply, and the handling of massive amounts of data. According to Navigant Research, worldwide MG generation capacity will increase to 7.6 GW by 2024 from 1.4 GW in 2015 [1]. Their alternating nature and inadequate storage capacity add a new layer of complexity to the task of PS operators maintaining power quality.

The majority of system functions are still performed manually or semi-automatic, leading to a shortage of AI approaches. As a result, there are still some obstacles and problems to overcome, like complex end-to-end control mechanisms as well as consumer involvement. In [24], safety analysis and fault detection of distributed generation and MGs were reported and recommended using inverters and ESS during operation. This research also discusses two MG operational options during an emergency: auxiliary energy source mode and inverter control mode. Kim et al. [25] investigated the benefits of an unconventional DES with a loop structure in voltage regulation and loss reduction. In addition, they provided a loop path prediction technique to minimize loss.

6.4.3 Integrating artificial intelligence into renewable energy sources

AI can stimulate the renewable energy sector, mainly through improved monitoring, operation, control, maintenance, and renewable energy storage, as well as more timely system operation control. The RES integration into the PS network has the following significant applications of AI [26]:

1. RES based energy generation, taking into interpretation the fluctuation of RES and the instability of supply;
2. Utility grid reliability, safety procedures and stability
3. Accurate weather prediction and load demand prediction
4. Effective DSM
5. An efficient way of energy storage option
6. Market planning and management
7. Improved grid component connectivity with MGs.

    Fig. 6.2 presents the relationships between AI and its applications.

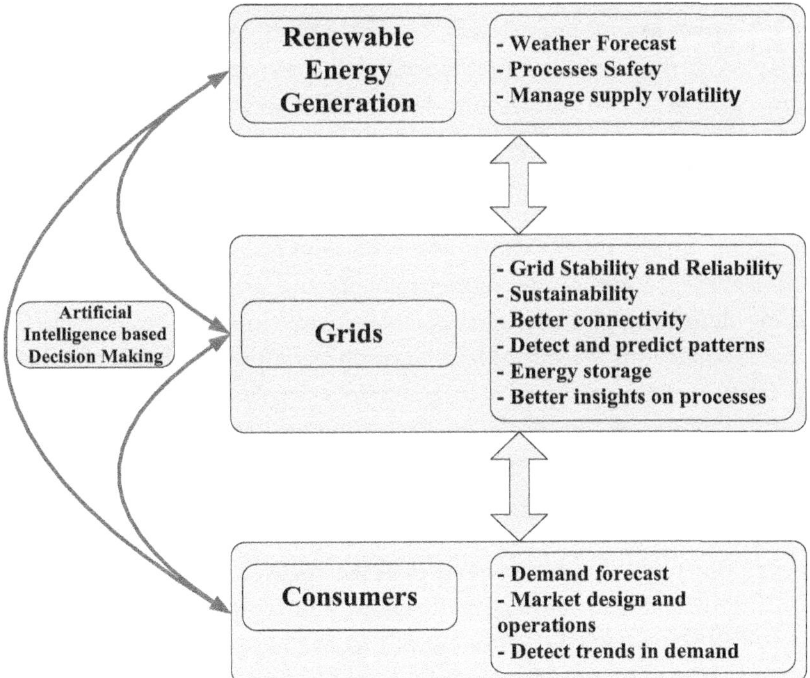

Figure 6.2 AI-based energy interaction. *AI,* Artificial intelligence.

In the renewable energy sector, effective uses of AI include intelligent supply-demand matching, intelligent storage, centralized control systems, and smart MGs.

6.4.3.1 Smart supply-demand matching

Even if the increased use of renewable energy offers today's societies a huge opportunity to tackle climate variation and resource shortage, there is a risk to renewable energy's ability to achieve the part of leader in the energy division, which is characterized by its intermittent nature [27]. AI might provide precise prediction in renewable based energy generation to prevent reliance on conventional power sources like fossil fuels, which allows operations to change without disrupting planning and responding to current customer demands. It is intended to boost the efficiency of renewable energy through automation procedures, an activity that would necessitate the widespread application of AI. The ultimate goal of the RES grid is to maximize generation capacity input while taking into account volatility and the associated uncertainty cost of the energy grid operation.

6.4.3.2 Storage intelligence

RE systems may become unstable if the RES does not have adequate energy storage size in the future due to variations in market density, virtual customers, load demand variations, and other factors. Recent advancements indicate that AI may provide optimization even in the absence of significant long-term meteorological data. By incorporating intelligent storage systems into renewable energy projects, which leads to optimize the return on investment (RoI) and enhance the response to varying demand and altering renewable energy inputs owing to changing climatic conditions. Renewable energy must be utilized to its most excellent capacity, and intelligent storage enables the development of smart RE system. Consequently, both energy providers and consumers will be affected, as they will have access to energy at a reduced-price level.

As transmission networks and storage technologies improve, they provide a considerable benefit for the integration of RES. Additionally, the flexibility to switch between different power sources is a beneficial adjusting mechanism. Storage technology can assist in resolving the volatility of RES, especially solar, wind, as well as the cyclical nature of demand. When there is no need for more energy, it is possible to create more of it.

6.4.3.3 Control system with centralized management

AI integrated into centralized control systems will help prevent scarcities in energy generation by identifying earlier problems. It will also assist to mitigate the amount of time required for maintenances. The system should have features such as reporting, statistics-based alarms, a backup server for recovery in the event of an unexpected failure, a user-friendly and web-based interface, security keys for verification of users from multiple places, and so on. The use of AI in RES is becoming increasingly necessary, given the increasing links between grids resulting from the vast information and data volume. The term "centralized intelligent control" means that the platforms used to manage supervised sites and facilities. AI could work this rapidly growing data, offer approaches for coping with renewable energy changes based on practice and prediction, as well as techniques for addressing variations in renewable energy.

There are a variety of assets with varying energy needs, technology, and ages. To assure their efficiency, they should be linked and reduced to a common denominator. According to the changing market conditions, the efficiency of a centralized intelligent unit should be confirmed by its ability to respond and adjust to the new requirements as quickly as possible, which is illustrated in Fig. 6.3.

6.4.3.4 Intelligent microgrids

MGs and smart incorporation into the utility power grid become more prevalent as their energy contributes to a more significant proportion of overall energy generation. In response to the increased interconnection between utility grids, MGs have transitioned from being independent systems (island) to centralized and regulated systems, and vice versa. Occasionally, MGs serve as the only technical alternative in situations where connections to the primary grid are restricted, and their output is determined by scheduling energy storage. Additionally, they may be a more cost-effective alternative in some circumstances [28]. In such cases, hybrid energy generation systems that combine wind and solar energy are becoming a more popular option.

Due to the growing need for renewable energy, increasing storage capacities, and improved communication linkages, it is necessary to enhance further grid protection and control systems, offering the energy system greater flexibility and reliability. Because of current SGs, conventional protection and control methods are becoming obsolete as they get

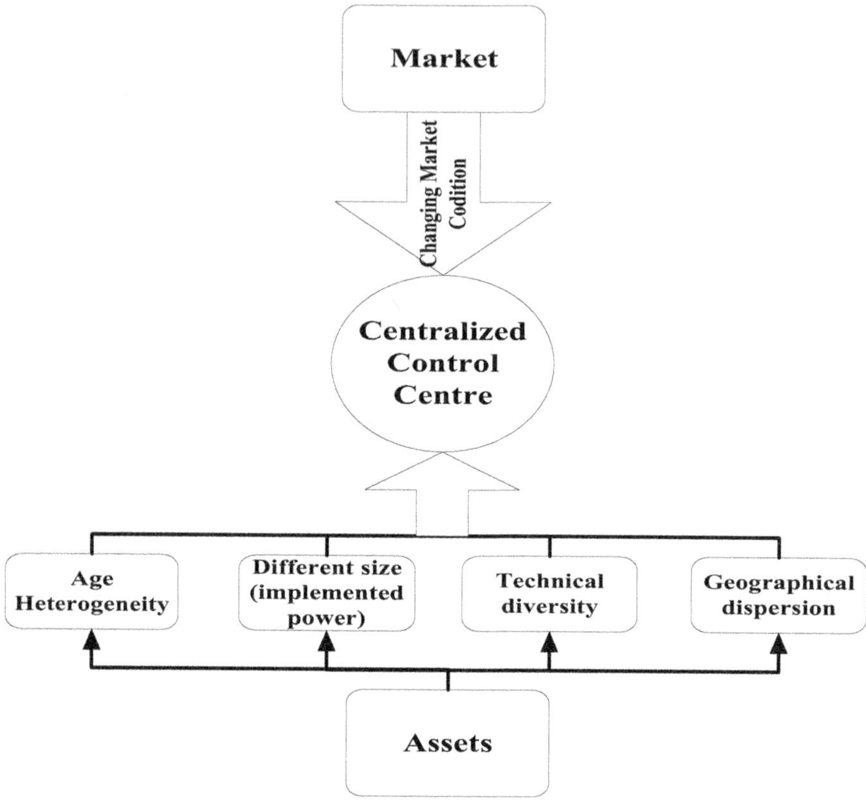

Figure 6.3 Centralized control center.

more complicated. The addition of "intelligence" to RES will boost the value of MGs through the following mechanisms:

1. Carry out the forecasting analysis
2. Improving the supply-demand balance
3. Increasing the production revenues and minimizing the time and cost of storage in the event of a quick change in direction for an unforeseen area of high energy demand
4. Storage and distribution in real time.

The synchronization between fluctuations in renewable energy generation and demand is difficult. Therefore, controlling energy generation, distribution, and energy storage while incorporating intelligence using AI techniques throughout the entire renewable energy generation and distribution chain will be essential to achieving a solution.

The successful integration of renewable energy and AI will be a game-changer for future SGs and will have a long-term influence on the industrial sector. The use of storage capabilities will aid in the adjustment to demand peaks. As a result, to reduce the cost of employing backup energy sources, the transition from delivering energy to customers from storage to generating power to energy users from generation should be accomplished as soon as feasible.

6.5 Artificial intelligence techniques for integration of energy storage system

An energy shortage has been a significant concern in the PS in recent times, which has directed to the retraction of the development of new energy resources. AI-based methodologies have been widely used in current systems to overcome the limitations of conventional methods. AI technology has been commonly used in the PS domain to predict energy demand, forecast PS failures, and provide reliable power. The AI-based system is also extensively used to indicate the power demand based on available ESS, which reduces the cost and complexity of the overall design. AI technology delivers power only to the critical load when there is a power shortage or outage. Table 6.1 illustrates the various AI methods utilized for ESS.

6.5.1 Energy storage systems integration

6.5.1.1 Energy storage system integration: consumer side

ESS is the future key component in SG aspects. ESS provides a reliable and uninterrupted power supply to consumers even during critical faults or outages. Future SGs will generally be equipped with RES. RES is typically intermittent, so ESS plays a vital role to ensure a continuous power supply. ESS also helps managing the power in the DG by reducing costs and enhancing its efficiency. ESS mitigates the peak load demand, which reduces the impact of establishing new power plants. It has been discussed in [34] how to implement inducement-based demand response programmes to persuade others to use such methodologies. Residential ESS stores electricity throughout off peak hours and delivers it through peak demand, falling the primary grid's reliability and providing financial aid to consumers. Among different ESS, batteries are the most promising technology because of their simplicity and wide application in electric cars and residential loads, thus reducing carbon emissions. By 2030, when ESS is fully integrated into the grid, it will provide electricity to 60 million

Table 6.1 Summary of artificial intelligence techniques utilized in the energy storage systems integration.

Reference	Year	Objective	Technique	Limitations
[29]	2018	Voltage and frequency forecasting at PCC	Artificial neural network	The stability of the system is not reliable
[30]	2017	Residential energy management system employing ESS	Optimization algorithm (PSO, GA)	Decrease in the peak to access ratio at the user end
[31]	2015	Reducing the cost of energy and annual loss	Sequential quadratic algorithm	Loss at the generation end and PCC not considered
[32]	2016	Storage sharing management	Optimization technique (Covex) and profit coefficient methodology	Dynamic load profile for consumers is not considered
[33]	2014	Power balancing technique for ESS and RES configuration	Lyapunov optimization	Communication of load and source is not considered for the power-sharing strategy

GA, genetic algorithm; *PCC*, point of common coupling; *PSO*, particle swarm optimization; *RES*, renewable energy sources.

people, contributing to a 30% reduction in carbon emissions from the power and transportation sector.

Integration of RES with ESS is the best suitable option to reduce fossil fuel-based energy generation dependence. Integration of ESS into RES will also provide a better alternative for providing uninterrupted power to consumers and storing excess capacity that can be delivered during peak demand. The ESS system is finding the most promising role in three aspects: reducing CO_2 emissions in generation and transportation, minimizing the dependency on fossil fuels by supporting RES-based power generation. ESS has an essential impact on SG operation in determining the economic dispatch solution at a market price [35]. Park et al. [36] presented a nanocirculation ESS based intelligent LED system to energize the city's streetlights, demonstrating ESS's importance in smart city initiatives. Energy storage sharing methodology can decrease the cost of consumers and also make the overall system compact. Rahbar et al. [32] developed an algorithm that uses the shared ESS concept to optimize the energy charged/discharged to benefit consumers.

6.5.1.2 Generation-side energy storage systems integration

A traditional power grid has a typical structure with minimized data and the possibility of energy storage, but a SG focuses on both aspects. The ESS is a vital component that has the ability to drastically modify the present grid's construction and operation. An intelligent power management methodology is needed to manage the dynamic properties of the DG to confirm efficient utilization of ESS [35]. The ESS has the ability to provide sufficient energy to all grid parameters at various levels, thereby increasing the reliability of the grid. Effective supervisory and economic policies for the growth and arrangement of ESS within a depository-based SG are essential for the successful incorporation of ESS to utility grid system as shown in Fig. 6.4. Thus, providing low-cost power to consumers and also providing effective market regulation for investors [37].

The reliability of the SG is improved by the distribution of forecasting technologies within the system. The issue of integrating ESS and the local low voltage DG at a point of common coupling is addressed in [29], which forecasts power factors using ANN technology. A real-time distributed system is discussed in [33] to satisfy the load demand through the ESS's charging characteristics. Using a nonsequential quadratic programming algorithm, a simultaneous optimization methodology for DG and ESS is discussed in [31].

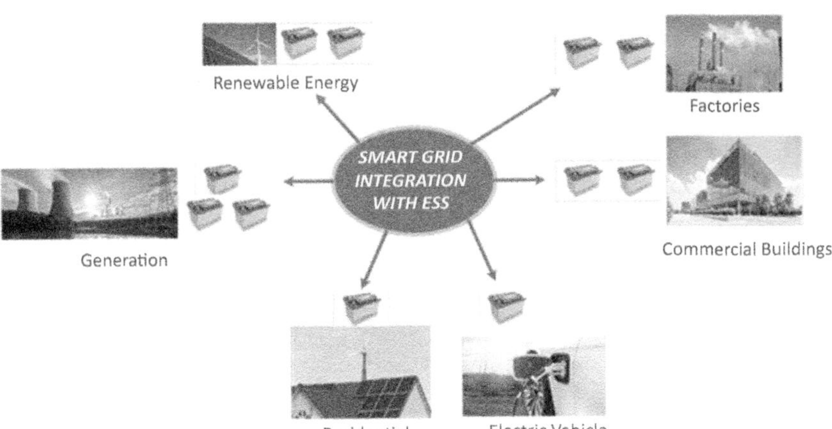

Figure 6.4 ESS integration with smart grid. *ESS*, Energy storage systems.

6.6 Economic feature and market deregulation in smart grid (current trends)

SG technology utilizes innovative commodities and facilities and emerging techniques for measuring, monitoring, control, and communication to integrate DG from RES. SG technologies creates a platform for the consumers to minimize their consumption, enhance their performance, encourage charging technologies for electric mobility, reduce their dependability on fossil fuels and increase their reliability. Fig. 6.5 determines the three major pillars of the SG infrastructure [38]. The SG technologies rely both on economic and sustainability aspects. The main principles of SG technology involve environmental protection, contributing to the reduction of CO_2 emissions, uninterrupted power supply, flexibility in managing peak demand, reliable distribution of electricity, and support for the deployment of RES, electric mobility, and better awareness of their consumption style for the end consumers.

SG technology systems can integrate multiple renewable energy systems into the system to provide an effective distributed system in low and medium distribution networks. This strategy has directed to the existence of a paradigm of generation and consumption of electricity, where customers consume power and produce it by installing small-scale wind and solar systems. The paradigms of production and consumption of power lead consumers to be prosumers. SG technologies need a comprehensive restructuring of conventional approaches to introduce new cutting-edge technologies to incorporate prosumers. The SG technology to involve prosumers in the distribution network is a complex process. Consumer behavior, culture, and practice have all been significantly influenced by the integration of generations and consumers throughout the entire value chain.

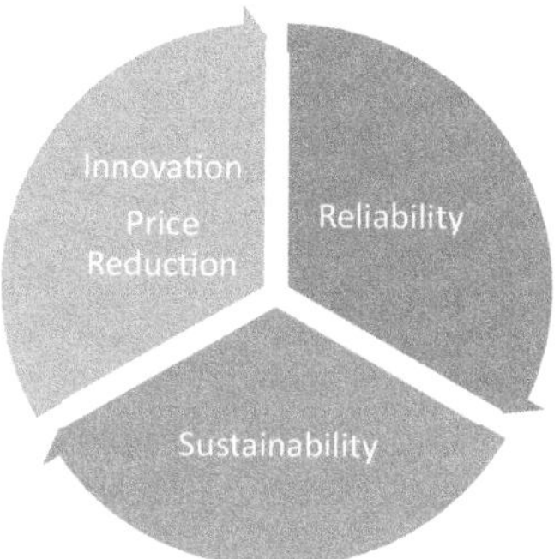

Figure 6.5 Smart grid infrastructure.

New socioeconomic aspects have been introduced, including the critically important issue of innovative ways to actively participate in the reduction of greenhouse gas emissions. Efficient SGs need the full cooperation of end users and residential users. Consumer acceptance, privacy, pricing, cybersecurity, and regulatory elements are socioeconomic constraints that are not analyzed by the impact of SG adoption [39]. As a result of these challenges, customers have been slow to accept new technologies in their intended form. To implement SGs globally, they must first understand how energy customers perceive the potential development of SGs and how their behavior changes as a result. Consumers' positive views of SGs as a solution to energy problems have been growing in recent years, but they are cautious of potential pricing rises. Consumers may be harmed by the complicated issue of SGs, which involves residential energy management and the actual installation of new devices in houses. Because of the complexity, people are unsure of the consequences of their decisions. The issue of customer privacy is an ongoing challenge in terms of social acceptance of SGs. The biggest disadvantage of SG is the potential for a massive quantity of data to be accumulated from smart apparatus that requires safe access. Not only does the SG technology assess consumer energy use, but it also detects their reliance on timing. Consumer data on energy usage, which defines their way of life, must be safeguarded, and data collecting is the most important technique to monitor power demand and apply dynamic pricing [40]. Data on the quantity of energy consumed can be utilized to monitor the consumer's behavior and its consequences.

The use of smart meters within the home creates the expectation that residents' activities will remain private. It is possible that sensitive data about consumers will be used for illegal or commercial purposes, by family members, by other parties for legal purposes, by law enforcement agencies, and other cohabitants. But it is related to the security measures in place to transfer information about individual customers to the energy supplier directly. The high initial costs of SG implementation can be a deterrent to the widespread adoption of SGs from a socioeconomic perspective. It is primarily the responsibility of institutional entities to provide a quantitative assessment of the costs associated with implementing SG systems.

Policymakers' role in establishing effective dynamic tariff plans that benefit the greatest number of energy consumers. SG technology may offer a solution in the form of a dynamic pricing structure to lower customer tariffs [41]. This reduces the need to install new power plants and enables consumers to alter their electricity usage patterns by minimizing consumption during the peak load condition. The existence of dynamic pricing methodology benefits is limited to monetary benefits and includes reduced electricity outage times and high usage of RES. Consumers' primary concern is the social/psychological danger of cybersecurity threats, which is a significant impediment to SG rollout.

Regulation is a critical role in the deployment of SGs. The climate of uncertainty acts as a barrier to delayed investment, incentives measures to enable investments. Statutory regulation influences market structure and the selection of RoI and organizational resolution. The expansion of SGs is mainly targeted at enabling public policies to be implemented for their implementation. SG technologies benefit the end-user by providing cost-effective and reliable power. Also, from an environmental point of view, it helps to mitigate challenges such as climate change and carbon emissions. Moreover, SGs contribute to rural development and allow energy independence; this reinforces the need for SGs in developing countries.

6.7 Challenges and issues for implementation of artificial intelligence-based distributed smart grid monitoring system

Conventional PSs are extremely complicated, and their control and analysis rely heavily on analytical and numerical computations. With the growth of SG with renewable energy and MGs, there are increasing uncertainties and challenges in the complicated environment. The conventional utility power grid becomes a SG system. Uncertainties are also added to modern SG systems because the current PS is based on outdated infrastructure. It's still a struggle for SGs to manage substantial amounts of data with significant fluctuation since the network is built on PSs.

In addition, researchers continue to work on the robustness, adaptability and online proceeding of AI systems [42]. Even though several data-driven methods for dealing with SG difficulties have been offered, there are still numerous significant hurdles. It is projected that the application of AI technology in SG will encounter the following significant obstacles based on the current state of AI technology development:

1. *Renewable energy integration:* The integration of RES has become increasingly important. High incorporation of renewable energy is a crucial attribute for the development of SGs. But there are several

significant issues because of renewable energy variability and unpredictability. Therefore, the output power can vary recurrently and abruptly [43].

2. *Data security and privacy preservation:* SG systems are more vulnerable to cyber–attacks than conventional power networks due to the utilization of different equipment and bidirectional connectivity. Numerous new safety measures have been developed to quickly identify cyber–attacks, system data theft, inserting false data, electricity theft, etc. However, the system is still exposed to many attacks using operating systems, network protocols, and physical devices in the existing SG. Current AI solutions for cyber–security with SG also compromise safety and performance.

3. *Demonstrate AI algorithms' learning capacity:* In general, AI algorithms suffer from the "black box" problem, meaning that they are incomprehensible and uninterpretable by human beings or other machines. This is an obstacle to AI algorithms.

4. *Inadequate data sample collection:* The application of massive data analysis in SG is still in its infant stage, and the amount of data collection in different application situations varies. Since data samples that match the requirements of various AI technology applications are insufficiently rich, the realization of AI applications based on small samples is a challenge that must be researched on an ongoing basis.

5. *Necessary to improve reliability:* However, even though AI applied to PSs has achieved a high degree of recognition rate for problems and fault findings, it has not yet met the needs of practical application. At this time, AI methods can only be employed as an auxiliary mode of operation.

6. *Necessity to upgrade infrastructure:* AI is based on the huge amount of data samples, distributed communication collaboration, and high-performance computation. But it is necessary to increase the supporting capacity and the number of key infrastructure resources, such as cloud computing, big data, and remote collaboration platforms, to achieve success.

7. *Shortage of power industry sector-based algorithms:* The algorithm is at the heart of AI technology. The algorithm adaptability of AI in a PS sector is currently limited compared to perception, prediction, and security maintenance. It is main aim of the subsequent study to recover the fundamental algorithm and provide a unique intelligent algorithm for PSs tailored to the industry's specific needs and characteristics.

6.8 Conclusion

In the face of climate change and environmental degradation, the SG and RES constitute a strong resource for future world development. To adapt to these specific requirements, AI entails new rules for managing processes. In this chapter, a comparative study of various AI techniques for supporting different applications in the future dispersed SG is discussed, including renewable energy integration, ESS integration, home energy management, demand response control, and protection. In addition to improving performance, these methods will simplify the process of managing SGs. AI approaches can be utilized to minimize the losses and enhance power quality in the distributing system. Furthermore, AI approaches can enhance and automate the control of dispersed resources, extend the range of SG activities, and facilitate the development of an even smarter grid.

References

[1] S.S. Ali, B.J. Choi, State-of-the-art artificial intelligence techniques for distributed smart grids: a review, Electronics 9 (2020) 1030. Available from: https://doi.org/10.3390/electronics9061030.

[2] U.S. Energy Information Administration — EIA — independent statistics and analysis. 2021. Annual Energy Outlook 2021. <https://www.eia.gov/outlooks/aeo/> (accessed 21.01.22).

[3] Smart Grid System Report — Department of Energy. 2022 <https://www.energy.gov/sites/prod/files/2019/02/f59/Smart%20Grid%20System%20Report%20November%202018_1.pdf> (accessed21.01.22).

[4] O.A. Omitaomu, H. Niu, Artificial intelligence techniques in smart grid: a survey, Smart Cities 4 (2021) 548–568. Available from: https://doi.org/10.3390/smartcities4020029.

[5] C. Ramos, C.-C. Liu, AI in power systems and energy markets, IEEE Intelligent Systems 26 (2011) 5–8. Available from: https://doi.org/10.1109/mis.2011.26.

[6] D. Neves, A. Pina, C.A. Silva, Comparison of different demand response optimization goals on an isolated microgrid, Sustainable Energy Technologies and Assessments 30 (2018) 209–215. Available from: https://doi.org/10.1016/j.seta.2018.10.006.

[7] G. Loganathan, D. Rajkumar, M. Vigneshwaran, R. Senthilkumar, An enhanced time effective particle swarm intelligence for the practical economic load dispatch, in: Proceedings of the IEEE Second International Conference on Electrical Energy Systems (ICEES) 2014. Available from: https://doi.org/10.1109/icees.2014.6924139.

[8] V.C. Gungor, D. Sahin, T. Kocak, S. Ergut, C. Buccella, C. Cecati, et al., A survey on smart grid potential applications and communication requirements, IEEE Transactions on Industrial Informatics 9 (2013) 28–42. Available from: https://doi.org/10.1109/tii.2012.2218253.

[9] R. Senthil Kumar, I. Gerald Christopher Raj, S. Sharavanan, Performance analysis of BLDC motor drive using enhanced neural based speed controller for electric vehicle applications, International Journal of Vehicle Structures and Systems 12 (2020). Available from: https://doi.org/10.4273/ijvss.12.2.24.

[10] Y. Saleem, N. Crespi, M.H. Rehmani, R. Copeland, Internet of things-aided smart grid: technologies, architectures, applications, prototypes, and future research directions, IEEE Access 7 (2019) 62962−63003. Available from: https://doi.org/10.1109/access.2019.2913984.

[11] S. Ben Slama, Prosumer in smart grids based on Intelligent edge computing: a review on artificial intelligence scheduling techniques, AIN Shams Engineering Journal 13 (2022) 101504. Available from: https://doi.org/10.1016/j.asej.2021.05.018.

[12] A.S. Khwaja, A. Anpalagan, M. Naeem, B. Venkatesh, Joint bagged-boosted artificial neural networks: using ensemble machine learning to improve short-term electricity load forecasting, Electric Power Systems Research 179 (2020) 106080. Available from: https://doi.org/10.1016/j.epsr.2019.106080.

[13] S. Vlahinic, D. Frankovic, B. Jurisa, Z. Zbunjak, Back up protection scheme for high impedance faults detection in transmission systems based on synchrophasor measurements, IEEE Transactions on Smart Grid 12 (2021) 1736−1746. Available from: https://doi.org/10.1109/tsg.2020.3031628.

[14] A. Mar, F. Pereira P, J. Martins, A survey on power grid faults and their origins: a contribution to improving power grid resilience, Energies 12 (2019) 4667. Available from: https://doi.org/10.3390/en12244667.

[15] S. Yoo, J. Eom, I. Han, Factors driving consumer involvement in energy consumption and energy-efficient purchasing behavior: evidence from Korean residential buildings, Sustainability 12 (2020) 5573. Available from: https://doi.org/10.3390/su12145573.

[16] Z. Liu, L. Wang, Leveraging network topology optimization to strengthen power grid resilience against cyber-physical attacks, IEEE Transactions on Smart Grid 12 (2021) 1552−1564. Available from: https://doi.org/10.1109/tsg.2020.3028123.

[17] E. Espe, V. Potdar, E. Chang, Prosumer communities and relationships in smart grids: a literature review, evolution and future directions, Energies 11 (2018) 2528. Available from: https://doi.org/10.3390/en11102528.

[18] Review of SCADA system for distribution power system automation, ERJ Engineering Research Journal, 42 (2019) 93−98. Available from: https://doi.org/10.21608/erjm.2019.66274.

[19] B. Eck, F. Fusco, R. Gormally, M. Purcell, S. Tirupathi, AI modeling and time-series forecasting systems for trading energy flexibility in distribution grids, in: Proceedings of the Tenth ACM International Conference on Future Energy Systems, 2019. Available from: https://doi.org/10.1145/3307772.3330158.

[20] M.N.Q. Macedo, J.J.M. Galo, L.A.L. de Almeida, C. de, A.C. Lima, Demand side management using artificial neural networks in a smart grid environment, Renewable and Sustainable Energy Reviews 41 (2015) 128−133. Available from: https://doi.org/10.1016/j.rser.2014.08.035.

[21] M.H. Elkazaz, A.A. Hoballah, A.M. Azmy, Operation optimization of distributed generation using artificial intelligent techniques, AIN Shams Engineering Journal 7 (2016) 855−866. Available from: https://doi.org/10.1016/j.asej.2016.01.008.

[22] N. Javaid, I. Ullah, M. Akbar, Z. Iqbal, F.A. Khan, N. Alrajeh, et al., An intelligent load management system with renewable energy integration for smart homes, IEEE Access 5 (2017) 13587−13600. Available from: https://doi.org/10.1109/access.2017.2715225.

[23] F.Y. Melhem, N. Moubayed, O. Grunder, Residential energy management in smart grid considering renewable energy sources and vehicle-to-grid integration, in: Proceedings of the IEEE Electrical Power and Energy Conference (EPEC) 2016. Available from: https://doi.org/10.1109/epec.2016.7771746.

[24] H. Jiayi, J. Chuanwen, X. Rong, A review on distributed energy resources and MicroGrid, Renewable and Sustainable Energy Reviews 12 (2008) 2472−2483. Available from: https://doi.org/10.1016/j.rser.2007.06.004.

[25] J.-C. Kim, S.-M. Cho, H.-S. Shin, Advanced power distribution system configuration for smart grid, IEEE Transactions on Smart Grid 4 (2013) 353−358. Available from: https://doi.org/10.1109/tsg.2012.2233771.

[26] A.C. Serban, M.D. Lytras, Artificial Intelligence for smart renewable energy sector in Europe—smart energy infrastructures for next generation smart cities, IEEE Access 8 (2020) 77364−77377. Available from: https://doi.org/10.1109/access.2020.2990123.

[27] E.B. Ssekulima, M.B. Anwar, A. Al Hinai, M.S. El Moursi, Wind speed and solar irradiance forecasting techniques for enhanced renewable energy integration with the grid: a review, IET Renewable Power Generation 10 (2016) 885−989. Available from: https://doi.org/10.1049/iet-rpg.2015.0477.

[28] M.D.A. Al-falahi, S.D.G. Jayasinghe, H. Enshaei, A review on recent size optimization methodologies for standalone solar and wind hybrid renewable energy system, Energy Conversion and Management 143 (2017) 252−274. Available from: https://doi.org/10.1016/j.enconman.2017.04.019.

[29] A. Massi Pavan, N. Chettibi, A. Mellit, T. Feehally, A.J. Forsyth, R. Todd, Ann-based grid voltage and frequency forecaster, The Journal of Engineering 2019 (2019) 3687−3691. Available from: https://doi.org/10.1049/joe.2018.8162.

[30] A. Ahmad, A. Khan, N. Javaid, H.M. Hussain, W. Abdul, A. Almogren, et al., An optimized home energy management system with integrated renewable energy and storage resources, Energies 10 (2017) 549. Available from: https://doi.org/10.3390/en10040549.

[31] E.E. Sfikas, Y.A. Katsigiannis, P.S. Georgilakis, Simultaneous capacity optimization of distributed generation and storage in medium voltage microgrids, International Journal of Electrical Power & Energy Systems 67 (2015) 101−113. Available from: https://doi.org/10.1016/j.ijepes.2014.11.009.

[32] K. Rahbar, M.R. Vedady Moghadam, S.K. Panda, T. Reindl, Shared energy storage management for renewable energy integration in smart grid, in: Proceedings of the IEEE Power & Energy Society Innovative Smart Grid Technologies Conference (ISGT) 2016. Available from: https://doi.org/10.1109/isgt.2016.7781230.

[33] S. Sun, M. Dong, B. Liang, Real-time power balancing in electric grids with distributed storage, IEEE Journal of Selected Topics in Signal Processing 8 (2014) 1167−1181. Available from: https://doi.org/10.1109/jstsp.2014.2333499.

[34] R. Deng, Z. Yang, M.-Y. Chow, J. Chen, A survey on demand response in smart grids: mathematical models and approaches, IEEE Transactions on Industrial Informatics 11 (2015) 570−582. Available from: https://doi.org/10.1109/tii.2015.2414719.

[35] A.H. Fathima, K. Palanisamy, Energy storage systems for energy management of renewables in distributed generation systems, Energy Management of Distributed Generation Systems (2016). Available from: https://doi.org/10.5772/62766.

[36] S. Park, B. Kang, Choi M-in, S. Jeon, S. Park, A micro-distributed ESS-based smart led streetlight system for intelligent demand management of the micro grid, Sustainable Cities and Society 39 (2018) 801−813. Available from: https://doi.org/10.1016/j.scs.2017.10.023.

[37] Z. Wang, C. Gu, F. Li, P. Bale, H. Sun, Active demand response using shared energy storage for household energy management, IEEE Transactions on Smart Grid 4 (2013) 1888−1897. Available from: https://doi.org/10.1109/tsg.2013.2258046.

[38] R. Kappagantu, S.A. Daniel, Challenges and issues of smart grid implementation: a case of Indian scenario, Journal of Electrical Systems and Information Technology 5 (2018) 453−467. Available from: https://doi.org/10.1016/j.jesit.2018.01.002.

[39] M.A. Ponce-Jara, E. Ruiz, R. Gil, E. Sancristóbal, C. Pérez-Molina, M. Castro, Smart grid: assessment of the past and present in developed and developing countries, Energy Strategy Reviews 18 (2017) 38−52. Available from: https://doi.org/10.1016/j.esr.2017.09.011.

[40] Social costs and benefits of Smart Grid Technologies. ISGAN 2019. <https://www.iea-isgan.org/social-costs-and-benefits-of-smart-grid-technologies/> (accessed 21.01.22).

[41] S. Bigerna, C. Andrea Bollino, S. Micheli, Overview of socio-economic issues for smart grids development, in: Proceedings of the Fourth International Conference on Smart Cities and Green ICT Systems, 2015. Available from: https://doi.org/10.5220/0005477402710276.

[42] M.S. Ibrahim, W. Dong, Q. Yang, Machine learning driven smart electric power systems: current trends and new perspectives, Applied Energy 272 (2020) 115237. Available from: https://doi.org/10.1016/j.apenergy.2020.115237.

[43] Y. Yoldaş, A. Önen, S.M. Muyeen, A.V. Vasilakos, İ. Alan, Enhancing smart grid with microgrids: challenges and opportunities, Renewable and Sustainable Energy Reviews 72 (2017) 205−214. Available from: https://doi.org/10.1016/j.rser.2017.01.064.

CHAPTER 7

Smart power quality control measures

D. Karthika and D. Renuka Devi
Department of Computer Science, Patrician College of Arts & Science, Chennai, Tamil Nadu, India

7.1 Introduction

It is crucial to have a high-quality control supply when power buyers are being controlled. In addition to the recurrence, voltage, and waveform characteristics of a controlled supply, it includes its accessibility. The power supply has been determined to be of good quality if it maintains consistent voltage levels, recurrence, and has smooth, sinusoidal waveforms. In practice, be that as it may, moving control prerequisites, certain gear (at domestic, at the office, and within the industry), and lacks can cause the plot to wander from what it is planning to do. The quality of the control may be compromised by the taking after conditions: when the control supply is hindered (power outages or impedances), when the given voltage is lower or higher than the run of satisfactory sizes, or when the control system repeat shifts. In this case, sinusoidal waveforms are distorted when current and voltage are applied. In other words, control quality is defined as the deviation of recurrence, current, and voltage from the intended values. In addition, the deviation may be depicted by the actual waveform (Fig. 7.1). In more particular terms, Control Quality alludes to the degree to which a control is adjusted with the smooth operation of electrical hardware. Subsequently, it might be a degree of how well a control system handles the loads. The client defines a control quality issue as to any voltage, current, or recurrence error that fails in the system or damaged equipment.

There may be concerns about the quality of destination control systems among consumers of residential, commercial, and mechanical products. Owing to inadequate control quality, bulbs and electrical hardware may malfunction or fail to work altogether, which in the long run can lead to early disappointment. As delicate generation machinery becomes increasingly computerized, quality control becomes increasingly challenging.

Smart Energy and Electric Power Systems
DOI: https://doi.org/10.1016/B978-0-323-91664-6.00002-4

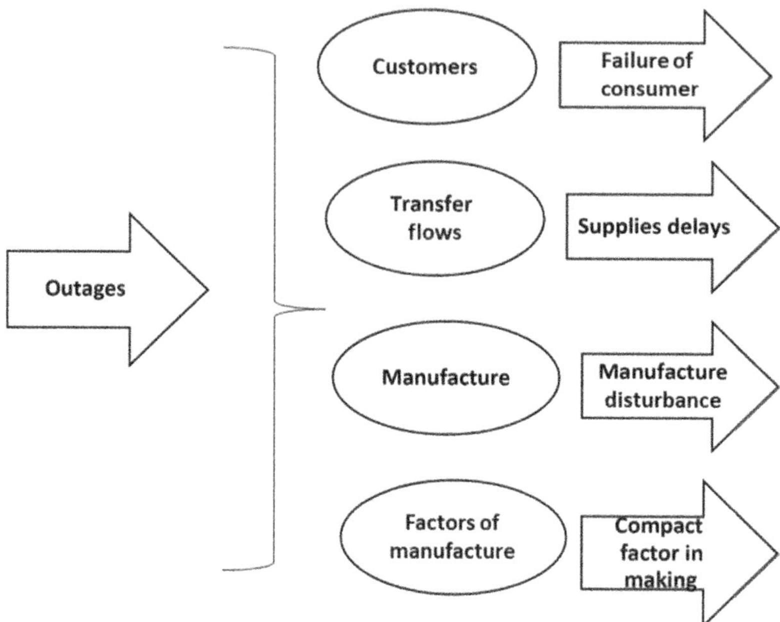

Figure 7.1 Causes of power quality.

Poor control quality can be resolved through a few causes and methods in most distributions. It may be possible to reduce such challenges. There are numerous applications in which control quality is an issue. These include circular segment heaters, airplane electrical frameworks, railroad frameworks, renewable energy, electric engines, mechanical devices, and electronic control transmission and dissemination frameworks. Control quality is determined by variables such as voltage frequency changes, temporal and acoustic characteristics, voltage, and current imbalances, etc.

7.2 Cause of reduced power eminence

When a control framework is transmitting electric vitality from the point of the era, a long removal from the stack centers to the point of utilization, with changing request, diverse climate conditions, uneven stacking on the scattering transformer stages, and clients utilizing modern and electronic hardware, numerous openings for a diminish within the transmitted control quality will exist. Control quality is compromised by exceptionally few issues. Changing control frameworks that alter the ostensible voltage value, change the sinusoidal waveform, or alter the recurrence stability will corrupt control quality. Examples are given in the following sections.

7.2.1 Voltage variety

Voltages are often problematic due to their extent. Voltage varieties can be characterized by a variety of shapes, and each shape is associated with a specific phrasing. A voltage variation occurs when a voltage value varies from its stated value for a short period (milliseconds to seconds) or a lengthy period (over 1 min). Variations of the voltage of short duration are generally manifested as plunges, droops, spikes, and swells, whereas those of long duration are manifested as flashes (voltage fluctuations), Undervoltage, overvoltage, and intrusions The voltage on the line fluctuate during these periods. Problems with the transmission or conveyance systems, exchanging capacitive loads and stacking cause voltage fluctuations.

7.2.2 Flicker

The giant effect is caused by arbitrary and tedious variations in voltage between 0.9 and 1.1 PV. It results from sudden changes in brightness and dimming of a screen and varying intensity of a light bulb's glow. Those effects are aggravating to humans. The output fluctuates as the electric engine runs, the throbbing stack, and the bend heaters are changed on and off.

7.2.3 Energy spears or surge power

Like voltage swells, surges are larger and last for a shorter period. Lighting strikes, arcing during circuit breaker and contactor switching operations, and surges or transients during switching are the most common causes.

7.2.4 Overvoltage

In this case, a heavy load has been turned off, transformer tap settings have been incorrect, voltage control has been inadequate, and a line has failed.

7.2.5 Undervoltage

A nominal voltage drop of more than 1 min below 0.9 pu is called an undervoltage. There can be several causes, such as switching on a high load or overloading a circuit.

7.2.6 Disruption

Interruptions that last more than a minute are considered persistent interruptions. A failure of the insulation, improper/faulty grounding, lightning, and insulator flashover are all possible causes. Using this technology, protective mechanisms open and close automatically, isolating the defective section of the system.

7.2.7 Outage

Power outages are caused when voltage is lost for a prolonged period. Falling on wires or poles, human errors, poorly coordinated systems, or a failure of protection measures [1].

7.2.8 Harmonics

A sound is produced by AC voltage and current, which are necessary by-products of the supply critical recurrence. An occasion consonant in a 50 Hz framework would be $2 \times 50 = 100$ Hz, a third consonant would be 3×50 Hz $= 150$ Hz, whereas the seventh consonant would be 350 Hz. A frequency that is not a number product of the control recurrence is an inter harmonic. A single mutilated waveform is produced when sounds and principal recurrences are combined. The odd consonants usually occur in a three-phase framework (3rd, 5th, 7th, etc.) [2]. Control quality issues are usually caused by consonant frequencies within the control frameworks. One of the main sources of control quality corruption in associated conveyance frameworks is the electric bend heater [3]. It generates noises, proliferates flashes, and causes fluctuations in flows and voltages [4]. A control conveyance framework consists of nonlinear stacks that produce harmonics. If a sinusoidal supply is used, their impedance will change with the provided voltage, and they will not draw sinusoidal current. An organization's voltage twists are created by the non-sinusoidal current because of the consonant substance that shapes the non-sinusoidal current [5]. There is a third consonant signal at 150 Hz.

7.2.9 Frequency fluctuation

Control framework recurrence variety or variance describes the deviation from the expected standard of operation (usually 50 or 60 Hz). When the control period is longer than the control request, the recurrence tends to drop. When the control period is shorter than the control request, the recurrence tends to rise. Blaming the transmission line, disengaging a huge stack, shutting down, or turning off a huge generator may also cause variances to recur. An external resilience value of * 5% isn't sound for a control system, as it could result in a collapse of the system.

7.2.10 Supply interruptions

Epilepsy control supply remains precarious, especially in developing nations. Quality control remains a major concern in these countries. It is caused by a deficiency in electricity demand within the network, which has arisen because a satisfactory venture is needed in the control division. Control quality problems are not insignificantly prompted by maturing control offices and a lack of support for existing ones. Issues with Control Frameworks and Case Studies 3−1 Changes within the broader environment that encompasses control frameworks and patterns reflected in control frameworks. There is a summary of the changes within the political environment and the structure of power in Japan, together with those in India, the EU part nations, and China, which is currently the world's biggest economy by GDP.

In India and China, control requests are increasing, whereas EU nations are implementing renewable energy and setting up DC frameworks to bind electric rate structures within the EU range. The interconnections between "distributed control sources" and "trade progression plans" (measures taken against blackouts) in the renewable energy generation are discussed in the following sections in the context of renewable energy, as well as a few cases of investigating these issues in the context of renewable energy generation. The sun-based and wind-based frameworks of control era systems are also dependent upon one another to function efficiently despite an unsettling control framework, and as a result, they contribute to the steady operation of control systems [2].

A momentary voltage drop may cause the acceptance machine of a wind turbine generator of coordinate lattice association sort to lose turn speed. The generator expands an increased amount of receptive control as a result; therefore there is a delay in the voltage recovery. By replicating precisely, the inertia and electric constants of the acceptance machine, it was possible to mimic the delay in voltage recuperation that is caused by a slip increase. To put it differently, expanding the capacity of the static VAR compensator (SVC), which is used to control voltage changes, will enable wind control systems to improve their voltage recovery characteristics. Is there any framework connection between conveyed control sources and issues that should be examined? [6].

While connecting, the most important things to remember are control sources and examples of how the measures should be implemented. In this segment, we examine a case of each thing under investigation. The following are two examples of these issues: concealment of voltage change inferred from yield change and avoidance of concurrent detachment in the event of a control framework unsettling effect. Whenever large-capacity-oriented frameworks are introduced in dispersion frameworks, it is indispensable to consider whether the yield voltage may change because of the variances of the control they produce. To control the fluctuations of voltage, wind control frameworks are generally equipped with SVCs.

7.3 Consequence of poor power quality

For the most part, control issues cause the voltage waveform of the supply to alter from sinusoidal to deviation from its apparent esteem or add up to a power outage. Issues with control quality can be last for milliseconds or hours. Electronic controls: private, mechanical, and office gear associated with an electric control supply may deliver electrically unsteady impacts coming about impacts, coming about in destitute control quality. Equipment such as scanners, computers, printers, etc., can create an electrical vibration that can devastate certain sensitive equipment related to the same supply source or in a few cases, cause them to glitch. Electrical unsettling influences are caused by mechanical drives fueled by electronic converters. The hardship of era happens when unsettling impacts happen or the quality of control is lacking, as appeared in Fig. 7.2.

The most notable impact of voltage droop is the early disappointment of hardware, loss of proficiency in turning machines, glitch of data technology gear, loss of information or solidity, handle hinder, glitch of measuring and control devices, etc. Voltage spikes and surges can cause electronic component damage, separation of separator materials, excessive screen brightness, harm to sensitive hardware, information preparing errors, information misfortune, and electromagnetic interference. Control waste is caused by sound, making control more wasteful and less effective in utilizing hardware. In this way, it affects the smooth operation of a mechanical machine, causing the generation to stop. In clinics, it can cause death. Various information handling exercises of information technology gear are affected, for instance, misplaced money exchanges in real-time [7]. Wiring and equipment overheat when consonants are present. Consonant frequencies interfered with the communication flag bring

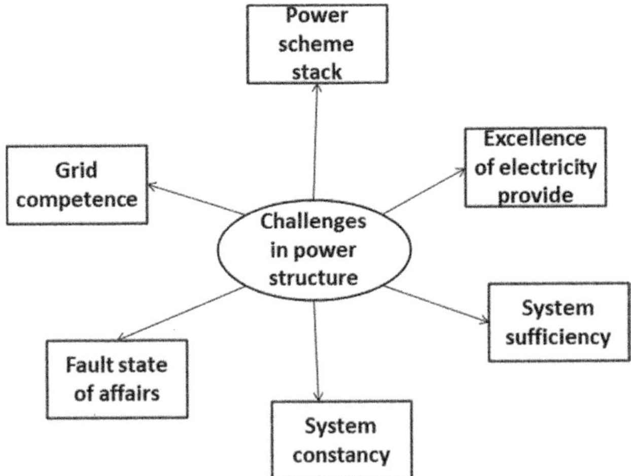

Figure 7.2 Effects of challenges in poor quality.

about incorrect communication flags when communication cables are parallel to control cables. It could lead to disaster. Consonant frequencies can affect defensive transfer operations inaccurately. Control Quality issues have taken a heavy financial toll. The additional fee a client must pay to avoid this bother may be analyzed in the form of an additional sum of cash.

7.3.1 Regulating standards on electrical quality

The most generally recognized benchmarks are the Universal Electrotechnical Commission and the Foundation of Electrical and Gadgets Design IEEE. These standards associations offer the least benchmark and satisfactory specialized hone and provide proposals on electrical and electronic specialized issues.

7.3.2 Power quality standard for equipment

The Computer & Commerce Gear Producers Association distributes ITI bends. Most Data Innovation Hardware ITEs can handle voltage runs (voltage list, swell, interference). While it was specifically constructed for the 120 V 60 Hz power supply, it can still be used with other voltage supplies (e.g., 240 V, 50 Hz) because it is scaled by the rate of voltage, not the magnitude. The ITI bend appears to be the point at which a piece of IT gear can work and withstand under voltages, overvoltage, and swelling at the rate of tensile voltage (vertical hub) [8].

After 25 cycles at 120% alleged voltage (swell), the overpowering stack will get swapped out. Gears can handle 200% of ostensible voltage for 0.05 cycles, then they go into denial. If the voltage is within the imperative performance mark, hardware can operate as it did for 0.2 ms or 0.01 cycles. When the voltage is too low, the gear may stop working or glitch. Voltage outside the shaded zone is shielded with intertwining or other defense devices.

7.3.3 Power quality monitoring

The control quality observing strategy is utilized to assemble, look at, and decipher rough control estimation data into important data [8]. Whereas the examination isn't restricted to fair these two sums, it incorporates planning for the estimation of the voltage and the current of the supply over a time outline and looking at their waveforms. Appraisals of wiring, setting up, and equipment associations are included. Control supply testing is fundamental for identifying current and potential control quality issues that abbreviate hardware's life expectancy. Quarterly plan (QP) checks are imperative for moving forward the quality of control in an office.

Control utilities are to guarantee that the quality of control given follows distinguished and reliable rules and to be arranged to standardize any specialized issues that influence the quality of control conveyed. Later progression in electronic and communication headways has opened the entryway to watching expansive and complex control systems effectively. In expansion to gathering data on an extended of control frameworks, utilities can assess the system and react appropriately as well as address complaints from their control clients [9]. Great control watching disobedience gives important information and a strong investigation of control quality. Within the outlines, there's a control screen that gives the voltage profile and waves shape of the supply voltage list, swells voltage assortment, and consonant level appraisal.

Within the Progressed Fault Recorder DFR, the activating occasion is activated when a fault happens and current, voltage, and waveform are recorded. The Unsettling impact can degree a wide extent of control unsettling impacts from a short-term passing voltage to long-term beneath voltages and control blackouts. The Streak Meter can be utilized to degree the level of voltage glare burden. Streak meters may be uncommon analyzers that show the response of a chain including a 60 W gleaming lamp-the eye-the brain of a typical onlooker. Having two critical parts, the primary parcel endeavors to reenact the operation of the lamp—eye—brain and the moment portion centers on measuring the transitory streak peculiarities [10,11]. The Circuit Screen gives exact, reliable, and speedy caution area and diverse levels of information on each control quality issue that can assist you to discover the source and cause of issues, counting consonant control streams, blazing, hanging, and swelling.

7.3.4 Mitigation technique

Awesome control quality movement can be guaranteed through a few measures. Adjustment of control quality issues can be conducted at different levels of the control system: at the control plant, at transmission lines and stations, at imperative and assistant conveyance frameworks, as well as at the advantage equipment and customers' building wiring. Even though the issue of control quality cannot be killed, many gears can be dispensed with, or on the off chance that a lightning strike can be maintained a strategic distance from or if fault can be dodged. The effect of control quality issues can for the most part be diminished to approach zero in any case.

7.3.5 Availability ensuring there is adequate power in the grid

The grid's ampleness alludes to the capacity of the control plant and transmission lines to meet the stack requests and vitality necessities of the clients. To meet client electric requests, the transmission, era, and dispersion framework inside the framework ought to be adequate [12,13]. Minimizing control quality issues depends on this.

7.3.6 Design of equipment

The producers of hardware ought to consider control quality concerns, and gear ought to be planned so that it does not contrarily affect control quality. Besides, the hardware ought to be planned to resist and be less vulnerable to control framework disturbances. By doing so, control quality issues can be minimized.

7.3.7 Interfacing devices

The provided connection can be interfaced with unstable equipment utilizing one of a few control electronic gadgets. The reason for this is often to avoid control quality issues inside the supply from coming to the hardware. In this case, we use a modified voltage controller AVR to preserve steady voltage into unstable equipment in any case of voltage hang, swell, and underneath or overvoltage. Another case is UPS gear which keeps equipment provided when there's a temporal control interruption. In expansion, the Lively Voltage Restorer DVR can be utilized to reestablish smooth sinusoidal line voltage when the source voltage waveform is debased or mangled. DVRs are voltage converters. DVRs are regularly utilized to interface between the control source and the control stack to be secured.

7.3.8 Filter

To avoid the undesirable banners from getting to the equipment, channels are utilized to permit the stream of required repeat. It is built with capacitors, inductors, and resistors to form a moo impedance channel for the fundamental repeat and a tall impedance channel for the repeat expected to be arranged. By employing a complementary consonant current, the consonant channel cancels the sound made by the nonlinear stack. There are different concordance channel sorts, such as energetic concordance channels, withdrawn concordance channels, line-reactors, electronic input channels, and uncommon transformers that utilize out-of-state windings to realize consonant lessening [14].

7.3.9 Proper grounding of the electrical system

In addition to protecting the installation, equipment, and users, good grounding can also improve the performance of the system. Poor power quality is often caused by inadequate earthing, particularly at the consumer end.

7.4 Future: opportunities and regulatory problems

In both the writing audit and own investigation, the expansion of transmission capacities has positive impacts. This framework would require high-voltage DC transmission lines that are cost-effective and effective. The investigation should be undertaken in this regard. The wind appeared to additionally be the favored innovation in the conditions and ought to be a center of investigation and improvement as well. To understand those frameworks, colossal capacities are not needed as adaptation occurs from interregional contrasts of era designs. Consequently, huge centralized baseload plants might in the distant future be atomic combinations [15].

A common reliance between nations, coupled with large-scale transmission framework expansions, may be a promising approach. It is perhaps the best choice from a technological and economic viewpoint, but its impediments arise from sociopolitical imperatives. The European Union faces challenges to maintain belief among its citizens at the moment. Having a healthy body is vital for the functioning and completion of nations, which in turn pushes patriotism in various countries. Despite this, more grounded participation appears to have numerous benefits for all nations participating. These benefits are particularly evident for nations facing extreme financial challenges. From end to end, global participation has many potential benefits, but political deterrents. Research and planning should strive to reveal these issues and build up positive energy participation.

Until now, participation in the EU has been hindered by national idiosyncrasies, while the ETS suffered the rot of costs, rendering it toothless. With these challenges overcome, the European arrangement offers a long-term, low-cost approach, which should energize European organizations to consolidate a European energy arrangement encompassing greenhouse gas emanation reduction, development of renewable energy sources, and building of a common foundation. The push for renewable energy sources and emissions reduction may prove to be an important European undertaking in the long run, which may at some point be as essential as the ECSC was more than 50 years back.

7.4.1 Identification of future challenges to power systems

Here are a few of the foremost pertinent challenges confronting future control frameworks. Point-by-point portrayals of each challenge set the arrangement for the definition of committed administration and control models and progressed capacities that can be connected to the distribution networks at the end [16]. See Fig. 7.3 for a rundown of these challenges. Connecting little conveyed vitality assets to conveyance grids.

Within the future, control grids ought to have a set of present-day highlights that will increase the execution, and which can fulfill the needs of the arranged chairmen and clients. A control system that considers the integration of unused emanant propels (electric versatility, microgeneration systems, and capacity systems). Microgeneration at the customer's premises, as well as the power delivered from loads and disseminated capacity units, with batte within the future, control grids ought to have a

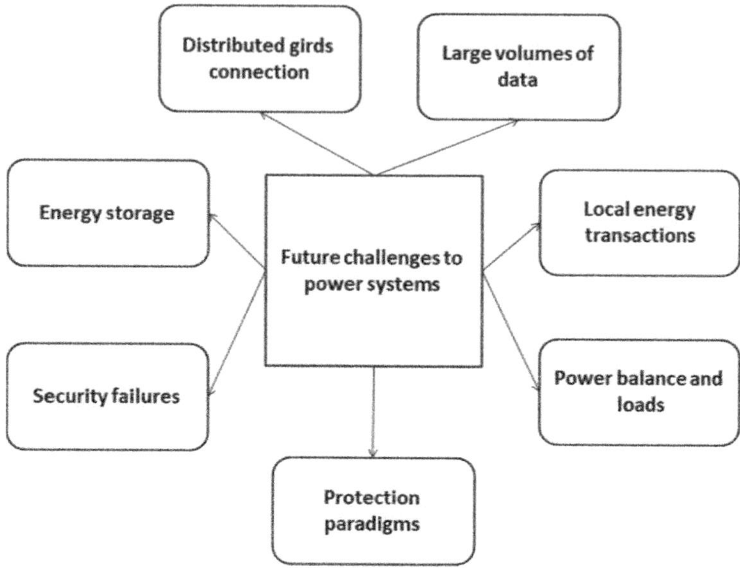

Figure 7.3 Future power grids.

set of present-day highlights that will increase the execution, and which can fulfill the wants of the organizing directors and clients. A control system that considers the integration of unused emanant propels (electric portability, microgeneration systems, and capacity systems). Microgeneration at the customer's premises, as well as the power delivered from loads and dispersed capacity units, with batteries from electric vehicles (EVs) working in vehicle-to-grid (V2G) mode, shape the bulk of DER. In the wake of consolidating these emanant developments, all of them creating (unidirectional or bidirectional) communication with the control cross-section, a strong control framework will be required to supervise the management trades in real-time, especially with the lively and passed on integration of electric transportability systems and ask response programs within the confront of lively charges. DG integration will require the allotment ofries from EVs working in V2G mode, shaping the bulk of DER.

7.4.2 Large volumes of data and heterogeneous data sources

Owing to the nonstop computerized change, an extraordinary bargain of data is being produced at numerous parts of the control section [17]. The data is of potential esteem to control companies, system directors, as well as to conclude customers and can be utilized through tremendous data calculations for an assortment of purposes such as evaluating the supply and request of vitality and developing bolster in vitality markets. It is expected that the data extricated from this data will decidedly influence operation and bolster costs, theory delays, and quality of advantage. In expansion, huge information preparing and mining are impelling the improvement of cutting-edge commerce models and expanding the accessibility of advanced (customized) vitality administrations.

Furthermore, in future control systems, it is expected that a noteworthy portion of the electrical machines will transmit in real-time to many components of control systems (e.g., framework repeat). These parameters are profitable for supervision calculations to survey whether the control signals are exact with the veritable operation of the control cross-sections in expansion to the basic control of electrical gadgets. A definition of rebellious and supervisory calculations is basic to dodging hurts inside the operation of the systems in case of disillusionments interior the command and organization systems or cases of ambush or intrusion within the communication framework, with the elemental deliberate of guaranteeing that electrical essentialness is made beneficially and effectively, both for buyers and control lattice directors alike.

7.4.3 Development of an end-user centric approach

It is expected that the electrical vitality will be utilized viably and productively within the future in an end-user-centric way, guaranteeing the dynamic involvement of clients together with the appreciation of the prerequisites of the control system. In this way, the advancement of continuously more beneficial, autonomous, and adaptable mechanical arrangements is required, allowing interaction between the different systems coupled to the control grids, conjointly communication systems permitting for the lively end-user affiliation. With such communication systems, it'll be conceivable to track utilization times for electrical gear for organize administrators and end-users, as well as join checking systems, to know in real-time vitality trades (for the occasion, when the client could be a buyer) and minimize inefficient and inconsequential utilization of the electrical gear. Additionally, by distinguishing such periods of operation for electrical devices, it can be conceivable to characterize control methods to optimize the utilization of the control arrange, for case, by utilizing stack moving, stack shedding, and thermostatic stack reparameterization disobedient, which incorporate beat shaving and stack leveling. With these rebellious, it is conceivable to progress coordination between ask and supply (i.e., inside a request-response framework). These courses of action incorporate M2M (machine-to-machine) supervisory control and information securing (SCADA) systems [14].

7.4.4 Local energy transactions

Additionally, by recognizing such periods of operation for electrical devices, it can be conceivable to characterize control strategies to optimize the utilization of the control organize, for case, by utilizing stack moving, stack shedding, and thermostatic stack reparameterization disobedient, which incorporate beat shaving and stack leveling. With these disobedient, it is conceivable to make strides in coordination between ask and supply (i.e., inside a request-response framework). These courses of action incorporate M2M supervisory control and information securing (SCADA) systems [14]. Several exhibit components are laid out for prosumers who have adjacent renewable vitality sources (RES) and who are willing to trade demand-side adaptability: (1) P2P exchanging for acquiring and advertising vitality organizations; (2) giving organizations to microgrids in utility coordination, and (3) providing organizations to a standalone microgrid. Cases of P2P trading stages.

7.4.5 High uncertainty in net load consumption balances and flows

The control systems of various countries have been able to suit rising offers of RES and this incline is set to continue. In any case, this float stance critical challenges to the existing control systems due to the inconvenience in adjusting with the basic time-varying nature of RES when these resources run the show time portfolios. Among other specialized challenges, repeat and voltage control can be influenced by RES-prompted bungles between supply and ask. In common, a positive relationship exists between the share of variable control sources and the control system flexibility needs. Flexibility, in this setting, concerns the degree to which such systems can supervise unpredicted working conditions, to be those started by the characteristic stack and scattered RES period changeability. Hence, without palatable flexibility components, it gets to be continuously harder to oversee with clutters between periods and ask stemming from their common assortments in honest to goodness time as the share of RES increments. In expansion, it to boot crucial to form unused organized organization devices that join information from vulnerabilities related to stack and adjacent time changes, such as stochastic OPF.

7.5 Conclusion

An in-depth analysis of the concept of power quality is provided in this study. In addition to identifying insufficient grid power as a source of power quality problems, the report also identified voltage fluctuations and deviations, frequency fluctuations, and waveform distortions in addition to inefficiency, overheating, and shortened equipment service lives. Although it is not possible to eliminate all the reasons for the power outage, the quality of the power supply can be improved, and the remaining influences on the power supply can be minimized or reduced. Several mitigation strategies have been developed, including using appropriate grid energy, using interface devices (such as UPSes, AVRs, DVRs, etc.), and enhancing power quality.

References

[1] R. Fehr, Harmonics made simple, Jan 2004 <http://ecmweb.com/archive/harmonics-made-simple>. (accessed 15.04.16).

[2] What are non-linear loads and why are they concerned today? Available at <http://www.mirusinternational.com/downloads/hmt_faq01.pdf> (assessed on 15.04.16), 2016.

[3] J. Mindykowski, Power Quality on Ships: Today and Tomorrow's Challenges. International Conference and Exposition on Electrical and Power Engineering (EPE2014), 16−18 October, Last, Romani, 2014.

[4] D. Bhonsle, R.B. Kelkar, Performance evaluation of composite filter fork power quality improvement of electric arc furnace distribution network, International Journal of Electrical Power & Energy Systems 79 (2016) 53−65.

[5] I.A. Ahmed, A.F. Zobaa, G.A. Taylor, Power quality issues of 3 MW direct-driven PMSG wind turbine, in: Proceedings of the Fiftieth International Universities Power Engineering Conference (UPEC), Sept. 2015.

[6] <http://www.selectricity.com/harmonicdistortionanalysis.html> (assessed on 15.04.16), 2016.

[7] T.G. More, P.R. Asabe, S. Chawda, Power quality issues and its mitigation techniques, International Journal of Engineering Research and Applications 4 (4) (2014) 170−177.

[8] A. Kusko, M.T. Thompson, Power Quality in Electrical Systems, McGraw-Hill, New York, 2007.

[9] P.V. Chopade, A. Brigade, D.G. Bharadwaj, Ensuring power quality and reliability − a step towards securing energy for the future, in: Proceedings of National Power System Conference NPSC, at Indian Institute of Technology (IIT) Roorkee, India, December 26−29, 2006, Page No:17−23.

[10] M.M. Canteli, Power quality monitoring, in: A. Moreno-Muñoz (Ed.), Power Quality: Mitigation Technologies in a Distributed Environment, Springer, London, 2007.

[11] J. Drápela, J. Šlezingr, A Light-flicker meter-Part I: Design, in: Proceedings of the 11th International Scientific Conference Electric Power Engineering 2010. Brno University of Technology, FEEC, DEPE, Czech Republic, 2010. pp.453−458.

[12] J. Popoola, A. Ponnle, T. Ale, Reliability Worth Assessment of Electric Power Utilityin Nigeria: Residential Customer Survey Results. School of Electrical and Information Engineering, University of Witwatersrand, Johannesburg, South Africa, 2011.

[13] D.O. Johnson, Reliability evaluation of 11/0.415 kV substations: a case study of substations in Ede Town, International Journal of Engineering Research & Technology 4 (09) (2015) 127−135.

[14] A. Agarwal, S. Kumar, S. Ali, A research review of power quality problems in electrical power system MIT international, Journal of Electrical and Instrumentation Engineering 2 (2) (2012) 88−93.

[15] R.P. Bingham, SAGs and SWELLs February 16, 1998. Available at: <http://dra-netz.com/wp-content/uploads/2014/02/sags-and-swells.pdf>.

[16] J.A.P. Lopes, A.G. Madureira, M. Matos, et al., The future of power systems: challenges, trends, and upcoming paradigms, WIREs Energy Environment e368 (2019).

[17] Y. Zheng, D.J. Hill, Z.Y. Dong, Multi-agent optimal allocation of energy storage systems in distribution systems, IEEE Transactions on Sustainable Energy 8 (4) (2017) 1715−1725.

CHAPTER 8

Directional overcurrent relay coordination optimization

Meng Yen Shih[1], Arturo Conde[2], Jorge Chan González[1] and Francisco Lezama Zárraga[1]
[1]Faculty of Engineering, Autonomous University of Campeche (UAC), Campeche, Mexico
[2]Faculty of Mechanical and Electrical Engineering, Autonomous University of Nuevo Leon (UANL), San Nicolás de los Garza, Mexico

8.1 Introduction

Protection elements are essential throughout the whole electrical network. Different robust protection principles have surged from around a century ago to offer protection to satisfy specific network purposes, such as stability, frequency, voltage, and thermal limits (targeted to offer reliability and continuity in the network). Also, these protection principles are employed in the different networks, whether in transmission, sub-transmission, distribution, meshed, ring, radial, active, passive, with or without renewable resources etc. Hence, protective relay plays a crucial role in the reliability, security, power quality, and lifetime of primary equipment.

Sub-transmission and distribution networks are mostly protected employing the "overcurrent protection principle." This is because overcurrent protection requires less investment compared to distance, differential, or pilot protection principles. Also, the overcurrent protection possesses the virtue of tolerating temporal overloading scenarios, which is common in distribution networks due to transformer inrush currents and motor starting currents. Moreover, overcurrent protection is capable to coordinate with downstream protection elements such as fuses and reclosers. These characteristics made the overcurrent protection the number one option for sub-transmission and distribution networks. However, a simple overcurrent protection can only be employed in radial distribution passive networks due to its incapability to determine fault position. Therefore, the directional overcurrent relay (DOCR) protection is employed in active distribution network, ring and meshed sub-transmission networks to discriminate faults and avoid false tripping (opening healthy circuits in its protection zone).

Smart Energy and Electric Power Systems
DOI: https://doi.org/10.1016/B978-0-323-91664-6.00011-5

The operating principle of DOCR is to monitor the present operating condition on the power network and operate protective breakers in the presence of a fault. This thermal threshold of the DOCR may offer protection precisely or in a roundabout way for transformers, buses, adjacent lines, and measuring equipment. The DOCRs can discriminate fault location (whether in front of or behind the DOCR) using the current and voltage phase relationship signals. Operation will not be executed if the fault is positioned behind the DOCR. However, if the fault is positioned ahead of the DOCR, then fault magnitude and reference pickup current will be compared to further determine if tripping signal should be sent to the circuit breaker.

Coordination must be analyzed and established among DOCRs to guarantee the sequence of primary and backup operations avoiding miscoordination and simultaneous trips. The DOCRs coordination study can be considered as an art and not an exact science, as the coordination process involves uncertainty to a certain degree, and it is also highly complex and nonlinear. Hence, global optimal result is normally difficult to yield and/or claim. Traditional coordination methods may lead to delayed operation time and curve intersections for large and/or meshed networks. Therefore, formulating the protection coordination problem to be solved with computational optimization methods is very much favored to obtain close to optimal results. Researchers have devoted their time to solve this problem, such as genetic algorithms (GAs) [1,2] to solve the coordination problem for meshed networks; particle swam optimization (PSO) [3−6] and its modified version; teaching learning based optimization (TLBO) [6−8] and its modified version; artificial bee colony (ABC) [9] algorithm; firefly algorithm (FA) [10−13] and its modified and chaotic versions; and hybrid algorithms [14−16].

8.2 Formulation of the coordination optimization problem

8.2.1 The objective function

The objective function (OF) is needed to coordinate the DOCRs using computational methods. This OF is the search engine to be optimized to yield close to optimal settings for the DOCRs. These settings must offer fast operations and correct primary-backup operating sequences. As the DOCRs need to offer primary and backup operation for the different circuits using the same protection settings, a robust OF need to be formulated. The OF is presented in Eq. (8.1):

$$\mathrm{OF} = f(x) = \left(\frac{\mathrm{NV}}{\mathrm{NCP}}\right) + \left(\frac{\sum\limits_{a=1}^{\mathrm{NCP}} t_{p_j}}{\mathrm{NCP}}\right) * \alpha + \left(\frac{\sum\limits_{b=1}^{\mathrm{NCP}} t_{b_i}}{\mathrm{NCP}}\right) * \beta + \left(\sum\limits_{L=1}^{\mathrm{NCP}} E_{\mathrm{CTI}_L}\right) * \delta$$

$$(8.1)$$

where t_{p_j} is the relay jth primary operation time, t_{b_i} is the relay ith backup operation time, α, β *and* δ are OF influence parameters, NV is the miscoordination accounting, NCP is the number of coordination pairs, and E_{CTI_L} is the Lth coordination pair CTI error.

8.2.2 The conventional inverse time relay characteristic

The inverse time relay curve according to IEEE standard C37.112−1996 [17] will be used in this chapter, displayed in Eq. (8.2):

$$t_i = \left[\frac{A}{\left(I_{\mathrm{sc}_{\max}^{3\varnothing}} / I_{\mathrm{pickup}_i}\right)^p - 1} + B\right] * \mathrm{dial}_i \qquad (8.2)$$

where $I_{\mathrm{sc}_{\max}^{3\varnothing}}$ is the 3-phase fault, A, B, p are the constants according to IEEE standard, t_i is the ith relay operation time, dial_i is the ith dial setting, and I_{pickup_i} is the ith pickup setting [18] accounting the maximum load profile to permit transitory overloading. A, B, p constants are presented in Table 8.1.

8.2.3 The dial and pickup setting constraints

The conventional parameter settings of DOCRs are namely the time dial setting and the pickup current setting, which are controlled within their superior and inferior cap. By doing so, optimized results are then

Table 8.1 ANSI/IEEE guideline for overcurrent protection.

Curve description	A	B	p
Moderate inverse	0.0515	0.114	0.02
Very inverse	19.61	0.491	2
Extremely inverse	28.2	0.1267	2

guaranteed to be suitable for DOCRs. Hence, the inequality constraints are displayed in Eqs. (8.3) and (8.4):

$$dial_i^{\min} \leq dial_i \leq dial_i^{\max}; \quad i = 1\ldots NR \tag{8.3}$$

$$I_{pickup_i^{min}} \leq I_{pickup_i} \leq I_{pickup_i^{max}}; \quad i = 1\ldots NR \tag{8.4}$$

where NR is the total relay number.

8.2.4 The coordination constraints

A coordination time interval (CTI) is designated for coordinating the DOCRs in longitudinal and meshed networks. The CTI parameters can ensure selective sequence for the relays. Thus avoiding sympathy or simultaneous relay operations. The CTI is designated previously as 0.2–0.3 s, so that the backup relay can intentionally delay its operation while not delaying too much if the primary relay fails to operate. The 0.3 s CTI contemplates a security limit, the relay disk over journey and breaker operation. Eq. (8.5) displays the CTI inequality constraint:

$$\text{CTI}_L \leq t_{b,i} - t_{p,j} \begin{cases} i = 1\ldots NR \\ j = 1\ldots NR \\ L = 1\ldots NCP \end{cases} \tag{8.5}$$

where CTI_L is designated as 0.3 s in this chapter.

8.3 The improved ant colony optimization

The ant colony optimization (ACO) algorithm is a family of swarm intelligence optimization method which was presented in 1992 by Dorigo. This method has been implemented successfully in different areas [19–24] and consists of an imitation of an ant colony searching for food, where ants leave pheromone in distinct paths and locations. Then, the optimized location and path are found when the shortest distance is encountered between the food source and the ant colony.

In real life, ants randomly hike around different places in search of food. An ant-k agent randomly drifts nearby its nest. Then when the agent come upon food, it goes back to its nest leaving pheromone on its path. So, when other ant agents come upon this path, they will follow this path to the food and then go back to its nest. Hence, this path will eventually be strengthened with more pheromone. If there shall be several

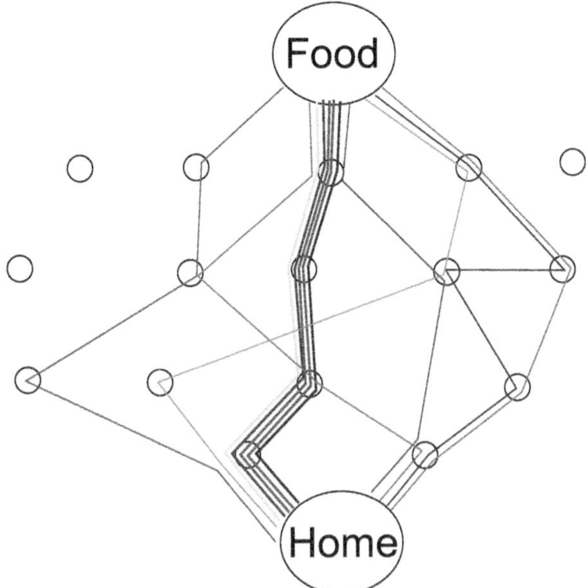

Figure 8.1 Ant colony optimization convergence.

paths to arrive home from the food, then as time goes on, the shortest path will experience more transit by ant agents compared to other lengthy paths. The pheromone on the shortest path will be increased and becomes greatly attractive. When time goes on, the lengthy paths will eventually vanish because of the pheromones' volatility. Then the ant agents have "determined" the shortest path. The idea is illustrated in Fig. 8.1 where the shortest path is greatly transited by the ant agents which has a higher concentration of pheromones. The ACO is composed of the following elements:

- Artificial ant agents are simulated to find results for the optimization problem where they transfer data of the paths' quality via concentration of pheromones. This emulates the behavior of real-world ants.
- The Ant Search Space (AS)-graph is the matrix search space which consists of discrete settings (states) of the control variables (stages). The matrix is a representation of the graph or sets of paths of the viable blends that the artificial ant agents will explore and examine.
- The pheromone matrix registers the concentration of pheromones left by the artificial ant agents regarding the different paths. Hence, the matrix reveals the attractiveness each discrete setting or possible paths to the solution. As more pheromone is deposited by the artificial ant

agents on certain setting, the greater the possibility it would be chosen by the artificial ant agents to solve the optimization problem.

- The transition rule is the instrument that the artificial ant agents employ to determine the concentration of pheromones and the feasibility of the settings. So that the artificial ant agent can select the following attracting location to inspect.

The pheromone update is executed to step up or step down the pheromone concentration for certain setting as reported by the artificial ant agent examination on whether the setting is a good or bad candidate to the solution. This may be accomplished by stepping up or down the pheromone rates through depositing or evaporating the pheromones, respectively.

At the beginning, numerous sets of solutions (states) within the feasible limits of settings are initialized. The AS-graph search space is formed by these sets of solutions, which is held fix during the entire optimization course (does not change from generation to generation). Whereas the pheromone matrix registers concentration of pheromone for every setting and the value of this matrix updates in every ant agent tour of each generation. Superior settings will have better fitness value and eventually all artificial ant agents would transit on this path and select these settings. Hence, the best result set is encountered.

This entire procedure is executed repeatedly until stopping criteria is met (maximum number of generations).

The number of states can be determined by the size of AS-graph. If there shall be very few states, then the ant agents will have lesser choices and capability to find close to optimal results (since only a small portion of search space will be examined). However, if there shall be too many states, then the ant agents will have greater choices and capability to find close to optimal results (since a big portion of search space will be examined but the optimization process may be slowed down exponentially).

8.4 Improved ant colony optimization for protection coordination

8.4.1 AS matrix

The AS-graph matrix, as mentioned previously, represents the search space with possible solutions to the problem. It consists of discrete settings (states) of the control variables (stages), where each discrete setting corresponds to a setting of the DOCR and each control variable corresponds to the number of different settings of the DOCR (freedom degrees).

$(m, n * \mathrm{NR})$ represents the size of AS matrix where NR represents number of relays, m represents number of discrete settings, n number of control variables.

For example, if the network under study has 15 DOCRs with 2 freedom degrees (*dial and k*), then the size of the AS matrix for a predetermined 25 discrete settings will be (25, 30) resulting 750 DOCRs discrete settings. As the number of discrete settings increases, the ACO may consume more time to examine and find optimized result. The AS matrix is presented in Eq. (8.6) in the following.

$$
\mathrm{AS} = \begin{bmatrix} dial_{(1,1)} & \cdots & dial_{(1,NR)} & k_{(1,NR+1)} & \cdots & k_{(1,NR*2)} \\ \vdots & \ddots & \vdots & \vdots & \ddots & \vdots \\ dial_{(m,1)} & \cdots & dial_{(m,NR)} & k_{(m,NR+1)} & \cdots & k_{(m,NR*2)} \end{bmatrix} \tag{8.6}
$$

The superior and inferior caps and the discrete steps of the control variables are essential to construct the AS matrix. For example, if the network under study has 2 relays with 2 freedom degrees (*dial, k*) where the DOCR parameter *dial* ranges in the interval [0.5, 1.4] of 0.1 step and the DOCR parameter *k* ranges in the interval [1.4, 1.6] of 0.05 step, then the AS matrix will be formed as presented in Fig. 8.2.

The two parameters have different step size and boundaries; hence, in this case, the *k* parameter has finished its discrete settings formation and just repeat the upper boundary of this parameter as illuminated in blue square.

$$
AS = \begin{array}{cccc} dial_{R1} & dial_{R2} & k_{R1} & k_{R2} \\ \end{array}
\begin{bmatrix}
0.5 & 0.5 & 1.40 & 1.40 \\
0.6 & 0.6 & 1.45 & 1.45 \\
0.7 & 0.7 & 1.50 & 1.50 \\
0.8 & 0.8 & 1.55 & 1.55 \\
0.9 & 0.9 & 1.60 & 1.60 \\
1.0 & 1.0 & 1.60 & 1.60 \\
1.1 & 1.1 & 1.60 & 1.60 \\
1.2 & 1.2 & 1.60 & 1.60 \\
1.3 & 1.3 & 1.60 & 1.60 \\
1.4 & 1.4 & 1.60 & 1.60
\end{bmatrix}
$$

Figure 8.2 AS matrix. *AS*, Ant Search Space.

8.4.2 Pheromone matrix

The pheromone matrix as mentioned previously, represents the concentration of pheromone left by the artificial ant agents. It reveals the quality of each possible path to the solution. Hence, as a discrete setting becomes more concentrated, it becomes more attractive to be selected by the artificial ant agents as their solution candidate.

The pheromone matrix $\gamma(m, n)$ has the same dimension as AS matrix. The matrix should be first initiated as shown in Eq. (8.7):

$$\gamma(m, n) = \gamma_0(m, n) = \tau_{\max} \tag{8.7}$$

where $\tau_{\max}$ presented in Eq. (8.8) is the maximum pheromone trail:

$$\tau_{\max} = \frac{1}{\alpha * f_{gbest}} \tag{8.8}$$

where f_{gbest} is the best overall result throughout the previous generations and α is a factor that is normally selected between [0.88, 0.99]. f_{gbest} may be assigned as the best initial guess solution to start the pheromone matrix when starting the optimization process.

Eq. (8.8) displayed how the pheromone matrix should be initiated with all equal edges when starting the optimization process. However, smallest relay settings are positioned in the first rows of the AS matrix according to Fig. 8.2. These first-row settings then become the most attractive relay settings to gain fast DOCR operation time. Therefore, the first-row pheromones are intentionally increased to help the algorithm to explore these paths leading to fast DOCR operation time and lesser computational execution convergence time. The engineer will have the freedom to increase or decrease the influence of first row pheromone matrix but should not alter in great quantity since algorithm exploration process may be stagnated. This concept is displayed in Fig. 8.3.

8.4.3 Transition rule

The ant agents use the stochastic-probabilistic tool (transition rule) to examine which are the tempting positions to visit next according to its previously deposited pheromone intensity. When ant-j is at the r-state of the $(i-1)$-stage, it will choose the s-state of the (i)-stage as the next visit according to the transition rule presented in Eq. (8.9):

$$p(r, s) = \frac{\gamma(r, s)}{\sum_l \gamma(r, l)} \quad s, l \in N_r^j \tag{8.9}$$

$$
\gamma_0 = \begin{bmatrix}
1.5(\tau_{max}) & 1.5(\tau_{max}) & 1.5(\tau_{max}) & 1.5(\tau_{max}) \\
1.5(\tau_{max}) & 1.5(\tau_{max}) & 1.5(\tau_{max}) & 1.5(\tau_{max}) \\
1.5(\tau_{max}) & 1.5(\tau_{max}) & 1.5(\tau_{max}) & 1.5(\tau_{max}) \\
\tau_{max} & \tau_{max} & \tau_{max} & \tau_{max} \\
\tau_{max} & \tau_{max} & \tau_{max} & \tau_{max} \\
\tau_{max} & \tau_{max} & \tau_{max} & \tau_{max} \\
\tau_{max} & \tau_{max} & \tau_{max} & \tau_{max} \\
\tau_{max} & \tau_{max} & \tau_{max} & \tau_{max} \\
\tau_{max} & \tau_{max} & \tau_{max} & \tau_{max} \\
\tau_{max} & \tau_{max} & \tau_{max} & \tau_{max}
\end{bmatrix}
$$

Figure 8.3 Pheromone matrix.

where N_r^j is ant-j memory list that records the not yet visited positions when the ant agent is positioned on location r. $\gamma(r, s)$ is the pheromone matrix that represents the probable next visit of (i)-stage presently under examination. $\sum_l \gamma(r, l)$ represents the pheromone sum of the whole column of the (i)-stage under examination. This concept is displayed in Fig. 8.4A−C.

In Fig. 8.4A−C, the light blue colors reveal the r-state of the ($i-1$)-stage and the pink colors reveals the s-state of the (i)-stage. Fig. 8.4 shows the ant agent finishing its journey in accordance with the transition rule. It is worth mention that the ant agents are randomly positioned to start the first column differently to avoid result stagnation and maximize algorithm exploration.

$$
\gamma = \begin{bmatrix}
0.8 & 1.5 & 0.7 & 0.5 \\
0.6 & 0.9 & 0.6 & 0.9 \\
0.7 & 0.7 & 0.9 & 0.7 \\
0.8 & 0.9 & 0.8 & 0.3 \\
0.9 & 1.1 & 1.2 & 0.3 \\
0.1 & 0.3 & 0.3 & 0.4 \\
0.4 & 0.1 & 0.5 & 0.1 \\
0.5 & 0.4 & 0.2 & 0.3 \\
0.8 & 0.5 & 0.1 & 0.3 \\
0.2 & 0.2 & 0.1 & 0.5
\end{bmatrix}
$$

(A)

$$
\gamma = \begin{bmatrix}
0.8 & 1.5 & 0.7 & 0.5 \\
0.6 & 0.9 & 0.6 & 0.9 \\
0.7 & 0.7 & 0.9 & 0.7 \\
0.8 & 0.9 & 0.8 & 0.3 \\
0.9 & 1.1 & 1.2 & 0.3 \\
0.1 & 0.3 & 0.3 & 0.4 \\
0.4 & 0.1 & 0.5 & 0.1 \\
0.5 & 0.4 & 0.2 & 0.3 \\
0.8 & 0.5 & 0.1 & 0.3 \\
0.2 & 0.2 & 0.1 & 0.5
\end{bmatrix}
$$

(B)

$$
\gamma = \begin{bmatrix}
0.8 & 1.5 & 0.7 & 0.5 \\
0.6 & 0.9 & 0.6 & 0.9 \\
0.7 & 0.7 & 0.9 & 0.7 \\
0.8 & 0.9 & 0.8 & 0.3 \\
0.9 & 1.1 & 1.2 & 0.3 \\
0.1 & 0.3 & 0.3 & 0.4 \\
0.4 & 0.1 & 0.5 & 0.1 \\
0.5 & 0.4 & 0.2 & 0.3 \\
0.8 & 0.5 & 0.1 & 0.3 \\
0.2 & 0.2 & 0.1 & 0.5
\end{bmatrix}
$$

(C)

Figure 8.4 Transition rule (A) Ant agent placed randomly on a state of first stage and perform transition rule to go to the following stage. (B) Ant agent perform transition rule on the previous state to go to a unknown state of following stage. (C) Ant agent perform the transition rule until reaching a certain state of the final stage.

8.4.4 Pheromone updates

The procedure of which the pheromone deposits are elevated or depressed in accordance with the examined results is called Pheromone Update. The pheromone deposits are elevated if good result has been found and depressed through pheromone evaporation when mediocre result has been found.

Local pheromone update: Each ant-j pheromone pathway journey is updated immediately as the ant-k agent finishes its journey. This is displayed in Eq. (8.10).

$$\gamma(r, s) = \alpha * \gamma(r, s) + \Delta\gamma^j(r, s) \tag{8.10}$$

$$0 < \alpha < 1 \tag{8.11}$$

$$\Delta\gamma^j(r, s) = \frac{1}{Q * f} \tag{8.12}$$

where α is the endurance of the pheromone pathway, $(1 - \alpha)$ describes the pheromone pathway evaporation and $\Delta\gamma^j(r, s)$ is the pheromone quantity that ant-j leaves on position(r, s). The appeal of position (r, s) is described by $\Delta\gamma^j$, which represents faster relay operation time.

The ant-j journey settings examination is described by f and Q is a positive factor. It can be seemed from Eq. (8.12) that as the factor Q elevates, the ant pheromone deposit depresses. Q has been chosen to be 100.

Global pheromone update: Primary and backup operation times are calculated after all ant agents have completed their journeys in the current generation. All pheromone positions (r, s) of the best ant journey are updated according to Eq. (8.13):

$$\gamma(r, s) = \alpha * \gamma(r, s) + \frac{R}{f_{\text{best}}} \quad r, s \in J^j_{\text{best}} \tag{8.13}$$

where f_{best} is the best result of the current generation, R is a positive factor and J^j_{best} is the location list of the best ant journey that records the state of each stage when ant-j moves from one stage to another. It is worth mentioning that as factor R elevates, ant pheromone deposits elevate as well. R has been chosen to be 5. The global pheromone update is executed randomly for some ant agents to avoid premature convergence and to obtain better exploration of the possible solution candidates.

8.4.5 Improved ant colony optimization

Improved ant colony optimization (IACO) is based on the idea to help ant agents explore and obtain better relay operation time after a determined number of generations. This is focused on the time dial setting only for certain relays, to comply coordination or to reduce operation time.

Initially, identify all the miscoordination relay pairs. Initiate with one mis-coordinated pair, from which the time dial setting of the primary relay is obtained as the superior cap limit using the best ant journey data. Then Eq. (8.14) is employed which contains a small adaption of Eq. (8.13).

$$\gamma(r, s) = \alpha * \gamma(ran1, s) + \frac{R}{f_{best}} \quad r, s \in J_{best}^{j} \tag{8.14}$$

where $ran1$ represents a selected random value between the span $[1{:}r]$. Then pheromone is deposited in this position. r is the state (superior cap) and s is the stage (specific relay). By performing the previous procedure, ant agents will be attracted to examine the respective setting from AS-graph. But since a superior cap is designated, the recently examined smaller dial settings will lead to faster primary relay operation time.

Afterwards, the time dial setting of the corresponding backup relay is obtained as the inferior cap limit using the best ant journey data. Then Eq. (8.15) is employed which contains a small adaption of Eq. (8.12).

$$\gamma(r, s) = \alpha * \gamma(ran2, s) + \frac{R}{f_{best}} \quad r, s \in J_{best}^{j} \tag{8.15}$$

where $ran2$ represents a selected random value between the span $[r{:}end\ of\ state]$. Then pheromone is deposited in this position. r is the state (inferior cap) and s is the stage (specific relay). By performing the previous procedure, ant agents will be attracted to examine the respective setting from AS-graph. But since an inferior cap is designated, the recently examined bigger dial settings will lead to slower backup relay operation time. Hence, the mis-coordinated pair of relays may become coordinated when the algorithm performs Eqs. (8.14) and (8.15).

On the other hand, shall there be zero miscoordination. Then, an improvement of relay operation time may be desired and executed. This initiate with identifying those relay pairs that may have big CTI values, then their respective time dial setting of these primary and backup relays should be obtained from the best ant journey, to be used as superior cap, to further optimized them. Afterwards, Eq. (8.11) is employed anew for both primary and backup relay. Then pheromone is deposited in these primary and backup relay positions. By performing the previous procedure, ant agents will be attracted to examine the respective setting from AS-graph. But since a superior cap is designated, the recently examined smaller dial settings will lead to faster primary and backup relay operation time. The flow diagram of DOCR optimization using ACO is presented in Fig. 8.5.

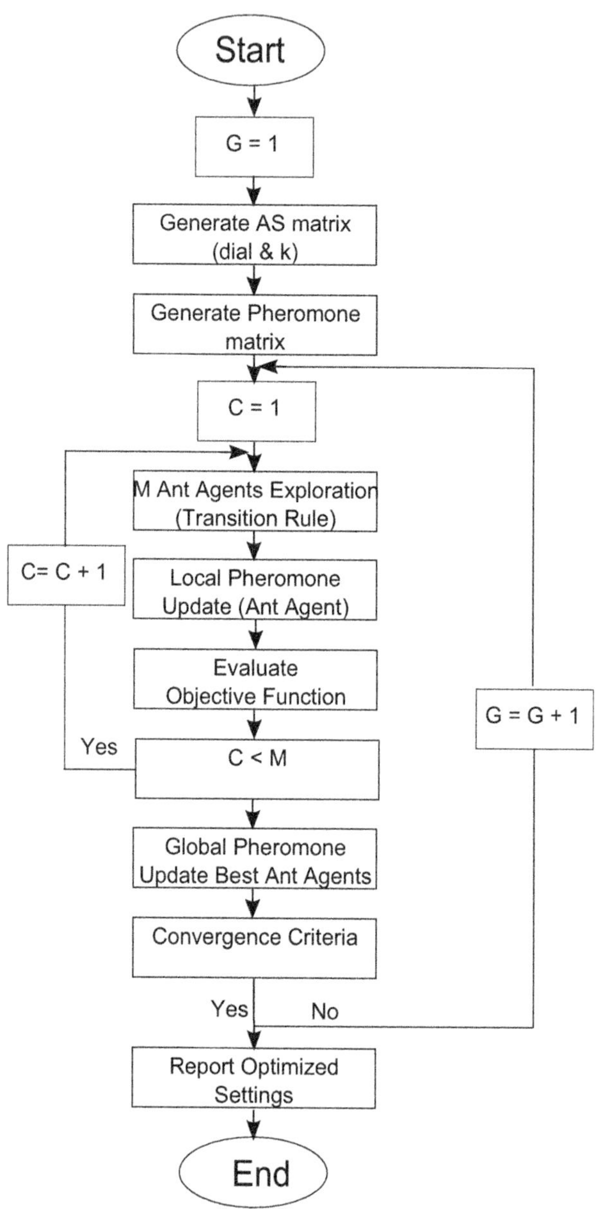

Figure 8.5 Flow diagram of DOCR optimization using ACO. *ACO*, Ant colony optimization; *DOCR*, directional overcurrent relay.

8.5 Improved differential evolution for protection coordination

Differential evolution (DE) algorithm [25] is a metaheuristic search evolutionary algorithm first reported as by R. Storn and K. V. Price in 1995.

The superiorities of DE can be summarized as:

- Simple and short algorithm which may be coded in a few lines compared to other evolutionary algorithms.
- Outstanding performances of algorithm robustness, convergence speed, and accuracy considering real world execution time.
- Very few control parameters, namely Cr, F and number of population (NP).
- Low space complexity, hence, it can lever large scale optimization problems.

The DE algorithm perturbs its actual generation individuals using scaled differences of different population individuals that are randomly chosen. Hence, additional probability operations are omitted for offspring creation. This made the algorithm very attractive and competitive since it requires less numerical process. Each individual of the DE algorithm is called parameter vector which encompasses a set of solution candidates.

The population is constructed by many parameter vectors or solution sets. Where the new population is constructed based on the old population using its selection procedure. This selection procedure will only replace the old population vectors if the new population vectors are better in fitness, hence, it will always get better results and never degrades. Then, the entire procedure is executed repeatedly until the convergence criteria is satisfied (maximum generation number reached etc.).

The size or NP reveals the quantity of parameter vectors in the population. The population size directly impacts the algorithm exploration process. If the NP increases, more solution candidates will be examined but execution time will be delayed. On the other hand, if the NP decreases, less solution candidates will be examined but execution time will be greatly enhanced.

The mutant scalar parameter F will aid exploitation is it is set near 0 and will aid exploration if is set near 1. The crossover scalar parameter Cr affects the algorithm to gain outstanding performance for numerous real world complex problems.

8.5.1 Initial population

The initial population consists of constructing a population of random initial conditions to be examined for the optimization process. Each parameter vector should be randomly initialized numerically within the DOCR's physical setting range. The rows of the population are called parameter vectors, and the columns represents the variables or settings of the DOCR. Hence, NP will have the dimension of $(NP, D * NR)$ where NP is the population size, D is the number of control variables and NR number of relays.

For example, if the network under study has 15 DOCRs with 2 freedom degrees (*dial* and *k*), the population size (NP) for 10 parameter vectors will be (10, 30) resulting in 300 variables or DOCRs settings. As NP increases the DE may consume more time to examine and find solution to the problem. The population (P) is presented in Eq. (8.16):

$$P = \begin{bmatrix} dial_{(1,1)} & \cdots & dial_{(1,NR)} & k_{(1,NR+1)} & \cdots & k_{(1,NR*2)} \\ \vdots & \ddots & \vdots & \vdots & \ddots & \vdots \\ dial_{(NP,1)} & \cdots & dial_{(NP,NR)} & k_{(NP,NR+1)} & \cdots & k_{(NP,NR*2)} \end{bmatrix}$$

$$(8.16)$$

The population should be randomly created between 0 and 1. Then Eq. (8.17) is applied to the population that has purely $0-1$ values, to start the initial population within its upper and lower boundaries.

$$P(p, q) = limit_{lower} + \left(limit_{upper} - limit_{lower} \right) * P(p, q)_{random} \qquad (8.17)$$

8.5.2 Mutation

The mutation process is a change or perturbation of all the present population parameter vectors with a random element executed in every generation. The target vectors or parents of the present population will be perturbed by a mutant vector called donor vector. Hence, the offspring or trial vector will be created by recombining the target and donor vectors together.

Three distinct vector index numbers should be randomly chosen from the DE population for each target vector. Pretend that the chosen parameter vectors are $\vec{X}_{r_1,G}$, $\vec{X}_{r_2,G}$, $\vec{X}_{r_3,G}$ for the *i*-th target vector $\vec{X}_{i,G}$. The r_1, r_2 and r_3 indices are independent and nonequal values from the span [1,NP], set up for every mutant vector. These values should also not be equal to the *i* index value. The difference vectors mutation scheme is displayed in Eq. (8.18):

$$\vec{V}_{i,G+1} = \vec{X}_{r_1} + F\left(\vec{X}_{r_2} - \vec{X}_{r_3}\right) \qquad (8.18)$$

where $\vec{V}_{i,G}$ represents the donor vector, F is the mutation factor usually within $[0.4,1]$, and $\vec{X}_{r_1,G}$, $\vec{X}_{r_2,G}$, $\vec{X}_{r_3,G}$ are the three randomly chosen parameter vectors.

8.5.3 Crossover

After constructing the donor vector via mutation, the crossover procedure come in play. This aims to intensify population diversification by swapping elements of the parent and donor vectors to construct trial vector $\vec{U}_{i,G} = [u_{1,i,G}, u_{2,i,G}, u_{3,i,G}, \ldots, u_{D,i,G}]$. DE crossover procedures are divided in two categories: uniform or *binomial* crossover and two-point modulo *or exponential* crossover.

Binomial crossover. This scheme performs a uniform recombination by randomly generating numbers in the interval $[0,1]$, then compare the random value with the scalar crossover Cr for every D variables. Hence, the trail vector may acquire roughly uniform distribution of parameter values from both the target and donor vectors. The binomial crossover scheme is presented in Eq. (8.19).

$$u_{j,i,G} = \begin{cases} v_{j,i,G}\, if\left(rand_{i,j}[0, 1] \leq Cr \text{or} j = j_{rand}\right) \\ x_{j,i,G}\ otherwise \end{cases} \tag{8.19}$$

where $rand_{i,j}[0, 1]$ is a random value that is generated uniformly. The random activity is carried out for every j-th element of the i-th individual. Afterwards, a randomly selected position $j_{rand} \in [1, 2, \ldots, D]$ assures that the trial vector $\vec{U}_{i,G}$ receives minimum one element from the donor vector $\vec{V}_{i,G}$. The scalar Cr is chosen as 0.5.

Exponential crossover. to perform this scheme, select an integer nn and L randomly between the values $[1, D]$. The value nn stands for the parent vector initial point where the elements swapping initiates. L denotes how many components the donor vector will transfer to the parent vector. Eq. (8.20) displays the exponential crossover.

$$u_{j,i,G} = \begin{cases} v_{j,i,G}\ if\left(j \geq (nn)_D \text{ and } L \geq L_c\right) \\ x_{j,i,G}\ for\ all\ other\ j \in [1, D] \end{cases} \tag{8.20}$$

where $(nn)_D$ is the starting point of crossover and L_c is the counter of L. which can be initially expressed as $L_c = 0$, then $L_c = L_c + 1$ for every evaluation of jth component; being that $L \leq D$.

8.5.4 Selection

The selection operation determines whether the trial or the target vector get through to the following generation, for example at generation $G + 1$. The selection operation is presented in Eq. (8.21).

$$\vec{X}_{i,G+1} = \begin{cases} \vec{U}_{i,G} \text{ if } f(\vec{U}_{i,G}) \leq f(\vec{X}_{i,G}) \\ \vec{X}_{i,G} \text{ if } f(\vec{U}_{i,G}) > f(\vec{X}_{i,G}) \end{cases} \tag{8.21}$$

here $f(\vec{X})$ represent the target vector fitness and $f(\vec{U})$ is the trail vector fitness. Whenever, a smaller fitness value is calculated from the recent trail vector, the target vector will be discharged and fill its place with the recent trail vector for the following iteration. If this case does not happen, then the target vector will be retained in the population. Therefore, by executing this procedure, the overall population will become better in fitness as generations goes on, or at least never degrades.

8.5.5 Improved differential evolution

The improved exploration of improved differential evolution (IDE) consists of renewing the three randomly distinct vector index numbers for each target vector. Since the original DE reported to determine these three indexes only once at the beginning of the algorithm, the proposed improvement is to renew these indexes every now and then. Hence, $\vec{X}_{r_1,G}$, $\vec{X}_{r_2,G}$, $\vec{X}_{r_3,G}$ for the i-th target vector $\vec{X}_{i,G}$ is continuously renewed in the optimization process. This greatly strengthened algorithm exploration and helped in converging at even better results. The flow diagram of DOCR optimization using DE is presented in Fig. 8.6.

8.6 Results performance study among genetic algorithm, IACO and IDE

Comparison results among GA, IACO and IDE are reported. Test systems has been selected to be the meshed IEEE 14 and 30 bus systems. The results are reliable for each algorithm since 15 simulation runs have been executed. It is worth mention that maximum and minimum load operating condition considering contingency analysis and the two double lines between buses 1 and 2 on the 14 bus system has been examined.

The formulation GA optimization for coordination is not presented. Since it is a widely known method and has already been reported in many literatures. Hence, it is used as comparison reference. IACO has

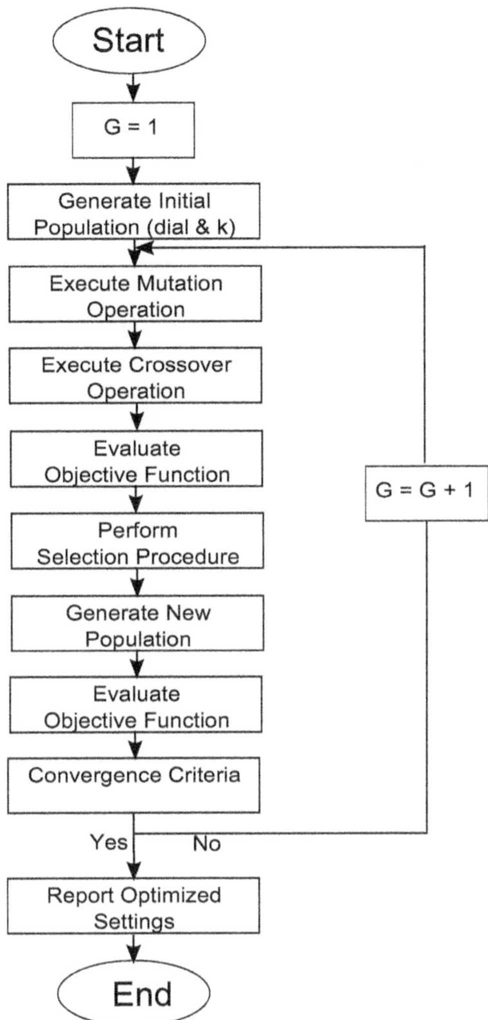

Figure 8.6 Flow diagram of DOCR optimization using DE. *DE*, Differential evolution; *DOCR*, directional overcurrent relay.

discrete *dial* and *k* settings whereas GA and IDE have continuous settings (Table 8.2).

8.6.1 IEEE 14 bus system

Table 8.3 shows the performance of the three algorithms at maximum and minimum load conditions. The presented results are averaged results of 1000 iteration of each simulation run, containing maximum, minimum

Table 8.2 Parameter settings of genetic algorithm (GA), IACO and IDE.

Parameters	GA	IACO	IDE
CTI	0.3	0.3	0.3
dial	[0.5:3.0]	[0.5:3.0]	[0.5:3.0]
k	[1.4:1.6]	[1.4:1.6]	[1.4:1.6]
dial step	Continuous	0.05	Continuous
k step	Continuous	0.01	Continuous
Q	—	100	—
R	—	5	—
Γ	—	—	0.5
F	—	—	0.8
Cr	—	—	0.5
Individual/Ant	500	500	500
Iterations	1000	1000	1000

IACO, improved ant colony optimization; *IDE*, improved differential evolution.

Table 8.3 Performance of genetic algorithm (GA), IACO and IDE for the IEEE 14 bus system based on execution time, fitness, number of violations and standard deviation.

Algorithm	t(s)	$f(x)$	NV	t-SD	$f(x)$-SD	$f(x)$-max	$f(x)$-min
Maximum load operating condition at 1000 iterations							
GA	2,344	4.72	0.00	106.30	0.18	4.92	4.35
IACO	255	3.53	0.00	187	0.33	4.10	3.07
IDE	42	2.25	0.00	0.56	0.01	2.26	2.23
Minimum load operating condition at 1000 iterations							
GA	2707	6.05	0.13	193.80	3.64	15.00	4.09
IACO	216	4.28	0.00	2.08	0.31	5.08	3.86
IDE	43	1.96	0.00	0.70	0.01	1.97	1.95

IACO, improved ant colony optimization; *IDE*, improved differential evolution.

and standard deviations. It is observed that GA used great execution time to accomplish the 1,000–iteration run. Whereas IACO and IDE managed to finish the simulations in lesser time. The averaged fitness values of Table 8.3 are presented in Fig. 8.7.

It can be seen from Fig. 8.7 that either for maximum load or minimum load condition, IDE has outperformed GA and IACO. Even though IACO converged at better fitness than GA, it is still a little poor compared

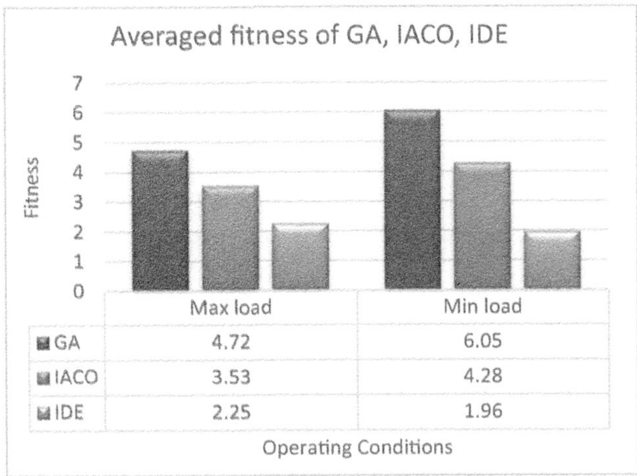

	Max load	Min load
GA	4.72	6.05
IACO	3.53	4.28
IDE	2.25	1.96

Figure 8.7 14 bus system: averaged fitness of GA, IACO and IDE in 15 runs. *GA*, Genetic algorithm; *IACO*, improved ant colony optimization; *IDE*, improved differential evolution.

Table 8.4 14 bus system: averaged primary, backup operation time and coordination time interval (CTI) of genetic algorithm (GA), IACO and IDE.

Algorithm	tp	tb	CTI
Maximum load operating condition			
GA	0.93	2.78	1.85
IACO	0.74	2.10	1.35
IDE	0.68	1.69	1.00
Minimum load operating condition			
GA	0.97	2.61	1.64
IACO	0.94	2.40	1.45
IDE	0.43	1.07	0.64

IACO, improved ant colony optimization; *IDE*, improved differential evolution.

to IDE. Since this is a minimization problem, the lesser the fitness value is, the better.

From Table 8.4, it can be observed that IACO has better results than GA for both maximum and minimum load condition. However, IDE has even better solution than IACO. Thus, IDE is concluded the best of the three algorithms for protection coordination problem.

An extract of results from Table 8.3 is presented in Fig. 8.8, where GA has prolonged execution times (2300 plus seconds) in different simulation runs. Whereas the IACO has less (200 plus seconds), and the IDE has occupied the least execution times (40 plus seconds) in the simulations. From their simulation time performances, IDE is preferred more than the other algorithms since it requires little time consumption.

Table 8.5 Detailed results of operation time and CTI of one simulation run using IDE on the IEEE 14 bus system.

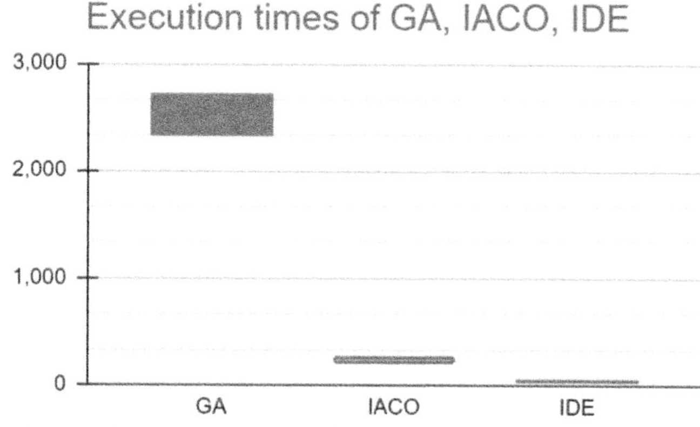

Figure 8.8 Averaged execution times of GA, IACO and IDE in 15 runs for the 14 bus system. *GA*, Genetic algorithm; *IACO*, improved ant colony optimization; *IDE*, improved differential evolution.

Table 8.5 Detailed Primary, backup operation time, and coordination time interval (CTI) of IDE for the 14 bus system for all coordination pairs in one simulation run.

Primary relay	Backup relay	IDE Min load		
		tp(s)	tb(s)	CTI(s)
2_1_2	1_2_1	0.2552	0.5970	0.3418
2_3_1	1_2_1	0.3270	0.6469	0.3199
2_4_1	1_2_1	0.2515	0.6455	0.3940
2_5_1	1_2_1	0.3399	0.6436	0.3037
1_2_2	2_1_1	0.2563	0.7183	0.4620
1_5_1	2_1_1	0.2462	0.8323	0.5861
2_1_1	1_2_2	0.2546	0.5959	0.3413
2_3_1	1_2_2	0.3270	0.6459	0.3189

(Continued)

Table 8.5 (Continued)

Primary relay	Backup relay	IDE Min load		
		tp(s)	tb(s)	CTI(s)
2_4_1	1_2_2	0.2515	0.6445	0.3930
2_5_1	1_2_2	0.3399	0.6426	0.3027
1_2_1	2_1_2	0.2574	0.7214	0.4640
1_5_1	2_1_2	0.2462	0.8361	0.5899
5_2_1	1_5_1	0.3046	0.7798	0.4752
5_4_1	1_5_1	0.4352	0.7813	0.3460
3_4_1	2_3_1	0.6467	0.9519	0.3052
4_3_1	2_4_1	0.3191	0.7097	0.3906
4_5_1	2_4_1	0.3835	0.6889	0.3054
5_1_1	2_5_1	0.3401	0.8034	0.4633
5_4_1	2_5_1	0.4352	0.7367	0.3015
2_1_1	5_2_1	0.2546	3.5150	3.2604
2_1_2	5_2_1	0.2552	3.5150	3.2598
2_3_1	5_2_1	0.3270	2.9206	2.5936
4_5_1	3_4_1	0.3835	2.2717	1.8883
5_1_1	4_5_1	0.3401	0.6520	0.3118
4_2_1	5_4_1	0.3377	0.6390	0.3013
4_3_1	5_4_1	0.3191	0.6587	0.3396
11_10_1	6_11_1	0.6897	0.9974	0.3076
6_12_1	11_6_1	0.4671	1.0551	0.5880
6_13_1	11_6_1	0.6756	0.9829	0.3073
12_13_1	6_12_1	0.3192	0.6227	0.3034
6_13_1	12_6_1	0.6756	2.1845	1.5089
13_12_1	6_13_1	0.2619	0.8754	0.6135
13_14_1	6_13_1	0.6057	0.9088	0.3031
6_11_1	13_6_1	0.8614	1.5264	0.6651
6_12_1	13_6_1	0.4671	1.7145	1.2474
10_11_1	9_10_1	0.7393	1.0433	0.3040
9_14_1	10_9_1	0.4944	0.9514	0.4571
14_13_1	9_14_1	0.4303	0.7402	0.3098
9_10_1	14_9_1	0.9212	1.2254	0.3042
11_6_1	10_11_1	0.6021	0.9043	0.3022
10_9_1	11_10_1	0.5734	0.8735	0.3002
13_14_1	12_13_1	0.6057	0.9152	0.3095
14_9_1	13_14_1	0.5281	0.8434	0.3154
13_6_1	14_13_1	0.3658	0.6661	0.3003
13_12_1	14_13_1	0.2619	0.7875	0.5255
AVERAGE	**0.4218**	**1.0580**	**0.6363**	

IDE, improved differential evolution.

Table 8.6 Detailed dial and pickup current settings of IDE for the 14 bus system for all directional overcurrent relays (DOCRs) in one simulation run.

Relays	Settings Min load			
	dial	k	Iload(A)	Ipickup(A)
1_2_1	0.5212	1.4011	389.40	545.59
2_1_1	0.5006	1.4000	389.40	545.18
1_2_2	0.5189	1.4043	389.40	546.84
2_1_2	0.5017	1.4022	389.40	546.01
1_5_1	0.5000	1.4011	280.88	393.54
5_1_1	0.5004	1.4095	280.88	395.89
2_3_1	0.6561	1.4355	281.56	404.18
2_4_1	0.5047	1.5208	260.82	396.64
4_2_1	0.5006	1.4100	260.82	367.76
2_5_1	0.6848	1.4948	226.85	339.10
5_2_1	0.5002	1.4001	226.85	317.61
3_4_1	0.5005	1.4010	283.17	396.71
4_3_1	0.5004	1.4088	283.17	398.94
4_5_1	0.5000	1.4008	288.06	403.52
5_4_1	0.6449	1.4000	288.06	403.28
6_11_1	1.6385	1.4008	194.19	272.02
11_6_1	0.9570	1.4005	194.19	271.97
6_12_1	0.8743	1.5180	230.27	349.54
12_6_1	0.5001	1.4003	230.27	322.44
6_13_1	1.2071	1.4641	292.11	427.69
13_6_1	0.5002	1.4007	292.11	409.16
9_10_1	1.6128	1.4032	331.51	465.16
10_9_1	0.5001	1.4002	331.51	464.17
9_14_1	0.8985	1.4633	287.45	420.62
14_9_1	0.5000	1.4056	287.45	404.04
10_11_1	1.3175	1.4128	225.98	319.28
11_10_1	1.0653	1.4010	225.98	316.60
12_13_1	0.5462	1.4605	158.13	230.95
13_12_1	0.5000	1.4222	158.13	224.89
13_14_1	1.0950	1.5329	176.07	269.91
14_13_1	0.7205	1.4118	176.07	248.59
AVERAGE	**0.7248**	**1.4254**	—	—

IDE, improved differential evolution.

Table 8.6 Detailed dial and pickup current results of one simulation run using IDE on the IEEE 14 bus system.

The threshold shows the pre-established 0.3 s as CTI. It is shown from Fig. 8.9 that all the 45 coordination pairs comply with the

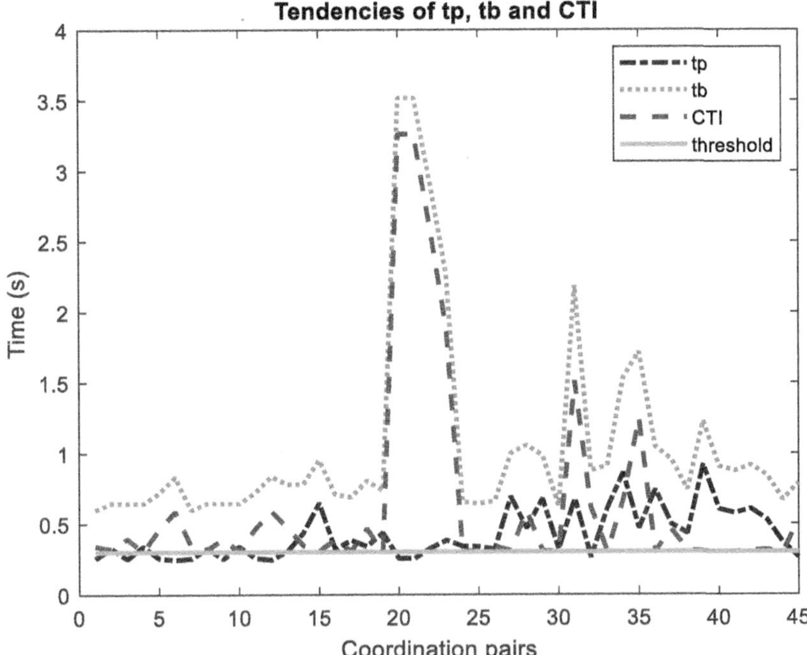

Figure 8.9 Tendencies of primary, backup and CTI times using IDE in one run for the 14 bus system. *CTI*, Coordination time interval; *IDE*, improved differential evolution.

optimization restrictions shown in Eq. 8.5. Even though some CTIs of certain coordination pairs have big CTI, the majority oscillate among the values 0.3—0.6 s. Hence, selectivity is maintained. The big CTI values for some coordination pairs are inevitable since the backup operation times have been increased, but the important aspect is to maintain selectivity, which has been accomplished. All primary, backup and CTI operation times are practical values for real life implementation.

Fig. 8.10 presents the convergence ability of IDE. This graph plots the best fitness value of each population in every iteration. The algorithm constantly works toward the best solution (minimization). The algorithm has almost converged at 500 iteration runs since the tendency of the fitness value has almost converged in a straight line and minimum slope has been observed.

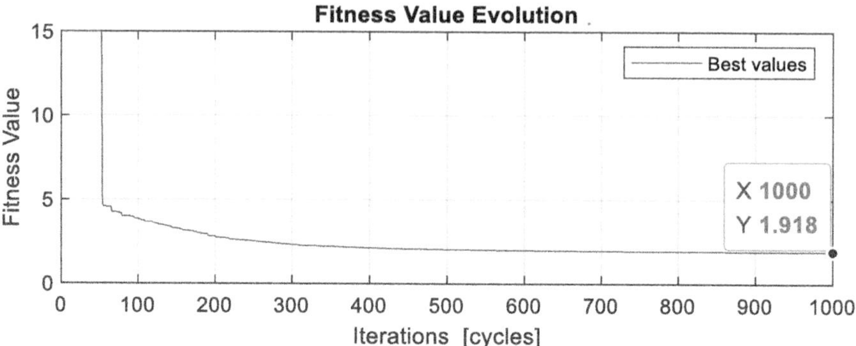

Figure 8.10 Fitness convergence of IDE in one run for the 14 bus system. *IDE, improved differential evolution.*

8.7 Summary

GA has been implemented in many different areas, where promising results have been reported. However, this well-known algorithm has shown limitations to handle the highly constraint and complicated DOCR coordination problem. The metaheuristic IACO has been proposed for this study, where execution time, fitness value, and relay operation times have been greatly improved compared to the GA. Nonetheless, IACO has been outperformed by the proposed IDE. This is observed from the maximum and minimum load condition studies on the IEEE 14 bus system where IDE has always obtained better solution sets than both the GA and IACO. One of the most impressive performances of IDE is the fast algorithm execution time (5 times faster than IACO and 60 times faster than GA) while obtaining better solution sets than both GA and IACO. It also shows better robustness and convergence ability as shown in Table 8.3, regardless of the random starting solution population and reporting zero miscoordination.

Acknowledgments

The authors are grateful for the support and effort of the Universidad Autónoma de Campeche and Universidad Autónoma de Nuevo Leon for its researchers to disseminate the research topics in which they are immersed.

Conflict of interest

None.

References

[1] M. Thakur, A. Kumar, Optimal coordination of directional over current relays using a modified real coded genetic algorithm: a comparative study, International Journal of Electrical Power & Energy Systems 82 (2016) 484—495. Nov.

[2] M.H. Marcolino, J.B. Leite, J.R.S. Mantovani, Optimal coordination of overcurrent directional and distance relays in meshed networks using genetic algorithm, IEEE Latin America Transactions 13 (9) (2015) 2975—2982. Sept.

[3] H. Zeineldin, E. El-Saadany, M. Salama, Optimal coordination of overcurrent relays using a modified particle swarm optimization, Electric Power Systems Research 76 (11) (2006) 988—995. Jul.

[4] M.M. Mansour, S.F. Mekhamer, N.E.-S. El-Kharbawe, A modified particle swarm optimizer for the coordination of directional overcurrent relays, IEEE Transactions on Power Delivery 22 (3) (2007) 1400—1410. Jul.

[5] M.N. Alam, B. Das, V. Pant, A comparative study of metaheuristic optimization approaches for directional overcurrent relays coordination, Electric Power Systems Research 128 (2015) 39—52. Nov.

[6] D. Saha, A. Datta, B.K. Saha, P. Das, A comparative study on the computation of directional overcurrent relay coordination in power systems using PSO and TLBO based optimization, Engineering Computations 33 (2) (2016) 603—621.

[7] M. Singh, B.K. Panigrahi, A.R. Abhyankar, Optimal coordination of directional over-current relays using teaching learning-based optimization (TLBO) algorithm, International Journal of Electrical Power & Energy Systems 50 (2013) 33—41. Sept.

[8] A.A. Kalage, N.D. Ghawghawe, Optimum coordination of directional overcurrent relays using modified adaptive teaching learning based optimization algorithm, Springer Intelligent Industrial Systems 2 (1) (2016) 55—71. Mar.

[9] M. Singh, B.K. Panigrahi, A.R. Abhyankar, Optimal coordination of electro-mechanical-based overcurrent relays using artificial bee colony algorithm, International Journal of Bio-Inspired Computation 5 (5) (2013) 267—280.

[10] R. Benabid, M. Zellagui, A. Chaghi, M. Boudour, Application of firefly algorithm for optimal directional overcurrent relays coordination in the presence of IFCL, International Journal of Intelligent Systems and Applications 6 (2) (2014) 44—53. Jan.

[11] M.H. Hussain, I. Musirin, A.F. Abidin, S.R.A. Rahim, Application of modified firefly algorithm for solving coordination problem of directional overcurrent relay, Journal of Electrical Systems (2015) 120—127. Dec.

[12] A. Tjahjono, D.O. Anggriawan, A.K. Faizin, A. Priyadi, M. Pujiantara, M.H. Purnomo, Optimal coordination of overcurrent relays in radial system with distributed generation using modified firefly algorithm, International Journal on Electrical Engineering and Informatics 7 (4) (2015) 691—710. Dec.

[13] S.S. Gokhale, V.S. Kale, An application of a tent map initiated chaotic firefly algorithm for optimal overcurrent relay coordination, International Journal of Electrical Power & Energy Systems 78 (2016) 336—342.

[14] J. Radosavijevic, M. Jevtic, Hybrid GSA-SQP algorithm for optimal coordination of directional overcurrent relays, IET Generation, Transmission & Distribution 10 (8) (2016) 1928—1937.

[15] A. Srivastava, J.M. Tripathi, S.R. Mohanty, B. Panda, Optimal over-current relay coordination with distributed generation using hybrid particle swarm optimization—gravitational search algorithm, Electric Power Components and Systems 44 (5) (2016) 506—517.

[16] A.S. Noghabi, J. Sadeh, H.R. Mashhadi, Considering different network topologies in optimal overcurrent relay coordination using a hybrid GA, IEEE Transactions on Power Delivery 24 (4) (2009) 1857—1863. Oct.

[17] IEEE Standard Inverse-Time Characteristic Equations for Overcurrent Relays, IEEE STD C37.112−1996.

[18] IEEE Guide for Protective Relay Applications to Distribution Lines, IEEE STD C37.230−2007.

[19] K.Y. Lee, J.G. Vlachogiannis, Optimization of power systems based on ant colony system algorithms: an overview, ISAP (2005) 22−35.

[20] Y.H. Hou, Y.W. Wu, L.J. Lu, X.Y. Xiong, Generalized ant colony optimization for economic dispatch of power systems, IEEE International Conference on Power System Technology 1 (2002) 225−229. Octuber.

[21] S.J. Huang, Enhancement of hydroelectric generation scheduling using ant colony system based optimization approaches, IEEE Transactions on Energy Conversion 16 (3) (2001) 296−301. September.

[22] T. Liao, K. Socha, A.M.M. de Oca, T. Stutzle, M. Dorigo, Ant colony optimization for mixed-variavle optimization problems, IEEE Transactions on Evolutionary Computation 18 (4) (2014) 503−518. August.

[23] M. Dorigo, L.M. Gambardella, Ant colony system: a cooperative learning approach to the traveling salesman problem, IEEE Transactions on Evolutionary Computation 1 (1) (1997) 53−66. April.

[24] M. Dorigo, M. Birattari, T. Stutzle, Ant colony optimization: artificial ants as a computational intelligence technique, IEEE Computational Intelligence Magazine 1 (4) (2006) 28−39. November.

[25] S. Das, P.N. Suganthan, Differential evolution: a survey of the state of the art, IEEE Transactions on Evolutionary Computation 15 (1) (2011) 4−31. Feb.

CHAPTER 9

Monitoring of wind power control in microgrid

Rajeswari Viswanathan
BVRIT HYDERABAD College of Engineering for Women, Hyderabad, Telangana, India

9.1 Introduction

The world's energy portfolio shows that the wind energy plays a vital role and designing and deploying of these wind turbines require the support of wind turbine control systems, which is again a primary necessity [1]. The main objective of conventional control systems for wind turbines is to increase the generation of energy by shielding the components of wind turbine. With respect to the increase in the generation of wind energy, the capacity of controlling the power output will increase. This satisfies the set points of power and regulates the frequency which is required for the utility grid. This may benefits the grid operators, since this may enable the wind turbines to become more productive compared to the conventional power plants. A broad outlook of a control system for wind mills [2] is discussed in this work. One of the important sources for the generation of electricity through renewable means is the windmill. Worldwide, more than 300 GW of electricity is generated from wind, which is improving thereby leading to 22% of them installed newly in the year 2018. Wind power's share of worldwide electricity usage at the end of 2018 was 4.8% from 3.1%. The power generation capacity in Europe was 18.8% in 2018. The capacity of wind energy is looked forward to grow continuously as a renewable energy source. Exceptional instrumentations are required to provide an accurate measurement of wind. Also, to evaluate the wind potential, continuous monitoring is needed before being deployment of the windmill since the wind is sporadic in nature. This work highlights developing a system to continuously observe and examine variables like speed of wind, direction of wind, wind pressure and temperature in the Lab VIEW environment wherein, the data acquisition system is used for the interconnecting. Also to estimate the values of power and power factor, the utility feeder's current and voltage is observed.

Smart Energy and Electric Power Systems
DOI: https://doi.org/10.1016/B978-0-323-91664-6.00007-3

The maintenance, monitoring, and assessment of prospective wind parks require world-wide standard measurement of wind and proper technology to be used [3]. So before deploying a wind park, accurate monitoring of wind is accomplished for proper assessment of the parameters of wind. Intermittent wind data analysis must be accomplished at least for a year and the data obtained thereby must be analyzed before deployment. A constant rate of energy will not be obtained from the energy conversion system, so the minimum energy available at the site must be calculated. For selecting the place for deployment of wind parks, few factors such as wind shift, turbulent conditions and velocity of wind etc., must be contemplated. Once the windmill is set up, accurate measurement of wind can be done during its operation. In this work, Lab VIEW is used to acquire, investigate and transmit the variables such as speed, direction, temperature, pressure of wind along with values of current and voltage in to a readable form. To get a precise assessment of wind resource, the present detailed wind data must be compared with the data from the meteorological department to clear off uncertainties. This can be accomplished to evaluate the wind power density of the place where it is to be deployed and also to check the suitability of the place of deployment. The global information says that the factors such as mean and annualized mean wind speed, wind power density, wind shear, turbulence intensity, mean air temperature, mean air density, speed frequency distribution, wind rose, daily and hourly distribution etc. are the vital parameters for assessing the wind resource. The power in the wind is directly proportional to the cube of the wind speed of the site, to the square of the diameter of the rotor and to the air density. It is also necessary that the energy produced annually and the production's incremental cost, etc., should be estimated for a wind turbine generator selected for the deployment. Based on the wind data obtained from the selected location, the blade of the rotor should be designed. These approximations determine the commercial successfulness of the deployment made. When the windmill starts operating, then it will be easy to analyze the performance of generator and intermittent observation assist in determining the robustness of the system. The response of the system with the availability of wind helps to identify the need for maintenance. Monitoring the velocity of wind is inevitable for regulating the gradient of the turbine above the rated speed of wind. One of the important characteristics of wind turbine is the power curve which shows the interrelation of power output and hub height wind speed [4]. The aim of providing power curve is to estimate the energy and to monitor the performance of turbines. Due to the expansion of windmill industry, the turbines are deployed under varied weather conditions, coastal areas, inshore, complicated landforms, etc., wherein these areas have maximum uncertainties which divert the curve from the standard values. Precise models of power curves will have an vital part in upgrading the performance of wind energy based systems.

This project showcases a thorough analysis of various perspectives of modeling the wind turbine power curve. The methods adopted to model rely on the motivation, data obtainability, and the expected outcomes with high accuracy. This also discusses about the purpose to model, several problems associated with it and also the accurate methods to be used for the measurement of power. The data used for modeling is derived from the Industry's specifications and the real time data from the wind mills. The main cause of concern in wind mills is the uncertainties caused by the nature of wind and the increased perforation of renewable energy in power systems. These may affect the reliability and stability of these systems. Precise models are needed for forecasting the desired output response and monitoring the performance of these systems, which even makes the system more reliable and stable. The mathematical relation showing the power taken by the turbine is

$$P = \frac{1}{2}\rho A_w C_p(\lambda, \beta)v^3$$

The speed, direction, and density are the parameters where the power production of wind depends on along with the parameters of the turbine. Moreover, it is highly complex to consider the consequences of all the parameters involved in this, which may lead to difficulties in evaluating the power output using the theoretical relation described earlier. This challenge can be overcome by using the Power curve for the wind turbine, since it showcases the power output of the turbine at a particular speed of wind. This in turn gives an appropriate method to estimate the efficiency of the wind turbines. Fig. 9.1 shows the graphical representation of power function of wind turbine. In this curve, the cut-in speed is the region where the speed of wind is lower than the minimum threshold value and the output power will be zero at this position. In the region 2, in-between the cut-in speed and the rated speed, a quick increment in power is noticed. In the region 3, the speed is constant till it reaches the cut-off speed. In region 4, above the cut-off speed, the operation of turbine stops to avoid complications for its components from high winds leading to zero power at this region.

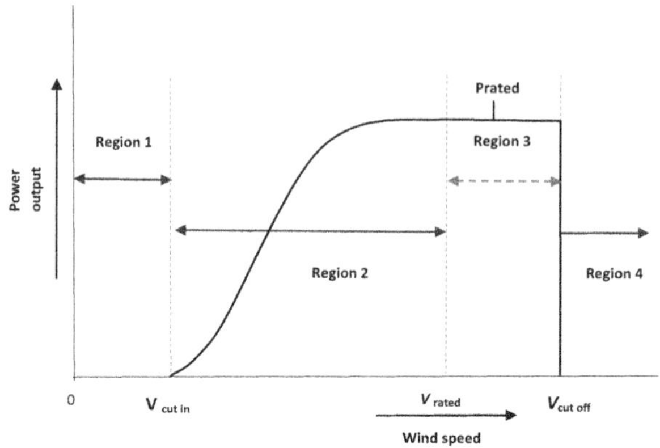

Figure 9.1 Graphic al representation of power function of wind turbine.

The objective of this project is mainly focused on three categories. They are

1. Monitoring of main grid.
2. Maintaining frequency constant.
3. Maintaining of output voltage constant.

9.1.1 Monitoring of wind turbine includes the following factors

1. Need for power curve modeling
2. Assessment of wind power and prediction
3. Capacity factor estimation
4. On-line supervision of power curves

9.1.1.1 Need for modeling of power curve

The response of wind turbine for various speeds of wind is shown by the power curve. Precise models of the power curves will be helpful for different applications of wind power generation.

9.1.1.2 Assessment of wind power and prediction

The assessment of wind power requires the details from the wind turbine power curve. To select the places suited for deploying wind mills requires detailed analysis of speed of wind, wind power density and potential of wind energy etc., of that particular region called as wind resource assessment. During this assessment process, the energy can be estimated using the data available and also from the power curve of wind turbine. While designing the wind energy based system, the size and cost involved to get the desired output must also be predicted. The power prediction made here should be accurate since overestimation or underestimation may lead to poor reliability or large structure of the system. Trading of energy to the electricity market also require proper forecasting of the power output to avoid inaccurate amount of power deliver to the market. Another alternative to model the curve is by using the real time data of wind speed derived from the data given by the SCADA system which provides better accuracy in predicting the power as it the real time data.

9.1.1.3 Capacity factor estimation

The ratio of the average output power to the rated power output gives the capacity factor of the wind turbine, and it also indicates the efficiency. This evaluates the average production of energy of a wind turbine that is needed to estimate the size and cost, optimum turbine-site matching, and ranking of potential sites. From the power curve models of wind turbine, the capacity factor can be calculated.

9.1.1.4 On-line supervision of power curves

The performance of wind turbines can be observed from the power curve. This power curve must require representing the performance of a turbine which operates under normal conditions which acts as reference. The output power measured and the wind speed data of wind turbines gives the reference curve. With respect to the reference curve, the actual curve of the turbine should be monitored. During this comparison, the diversion of the values of the actual curve compared to the expected output may represent the faults. The output of the wind turbine may be affected by the underperformance of the turbine such as blade faults and yaw and pitch system faults. Various faults affect the turbine system in a different way and this may lead the power curve to divert from the antici-pated value.

9.1.2 Maintaining constant frequency and voltage

A point-of-load pitch control is used. The input power and the power output should match provided the input power of windmill changes owing to the high speed of wind and also a load with a power demand is connected. When the power output is given owing to the load, then the wind generation unit should adapt input to output. Since the output power is related to torque, and if it is desired to provide a constant speed or frequency then it is mechanical rotational system. Pitch-control is used to switch the gradient of rotor vanes and to transmit the rotational force to the generator. Depending on the increase in wind speed, torque is produced thereby harvesting the wind energy, keeping speed time's torque-the power – constant at constant frequency.

9.2 Problem statement

In India, the deployment of electric grid with renewable energy has attained nearly 33,792 MW [5]. In the year 2015, out of the total capacity of 260.8 GW, the wind power was about11% and nearly 66.5% of entire capacity of renewable energy. With the increased perforation of wind power, it is more vital to ensure that the perforation of wind power does not influence the power quality, stability and reliability of the networks in power system under various conditions of operation. Due to less investment, good capacity to transfer energy and nature of adaptability, the variable speed wind turbines with doubly fed – induction generator (DFIG) are capturing much attention in the power system sectors compared to other technologies. Detailed study of the design and stability related issues should be analyzed prior to using the DFIG to the grid. And simulation can be done for proper modeling of the system. The objective of this work is to analyze the suitability of using a custom designed wind turbine and to monitor the wind power generated using a prototype model of wind turbine to improve the output garnered from the wind energy system. Fig. 9.2 gives us the detailed explanation of the DFIG wind turbine used in the project and its performance is evaluated.

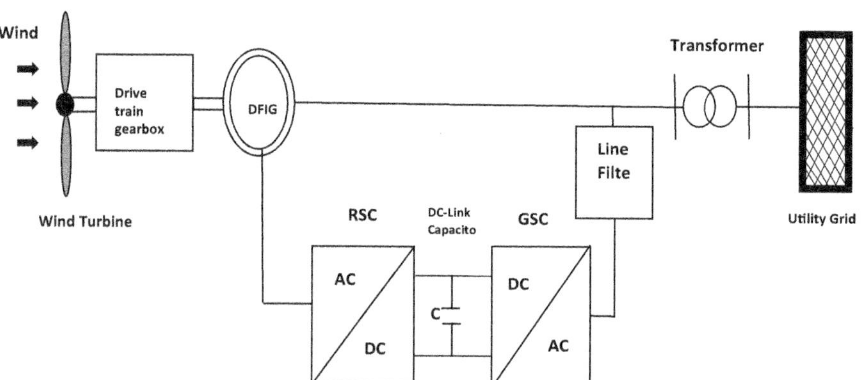

Figure 9.2 Schematic arrangement of DFIG based wind energy system.

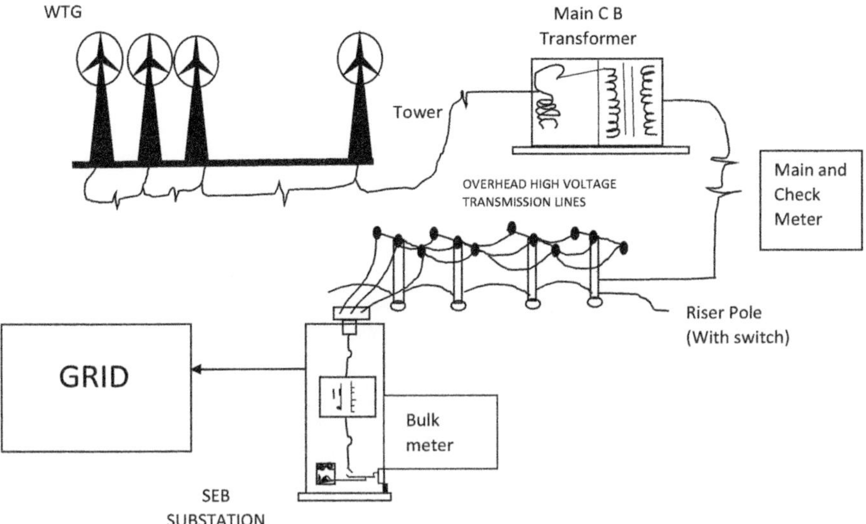

Figure 9.3 Wind energy farm methodology software design.

In the generation of wind energy, the electrical energy is obtained by converting the driving force of wind to mechanical energy. As wind blows at high speeds, the driving force will be high in wind. When the driving force strikes the vanes of wind turbine, it gets converted to mechanical energy and due to this torque developed, the vanes starts rotating. With the rotation of vanes, the rotor connected to it also starts rotating which in turn triggers the generator to produce the electrical energy. This conversion takes place due to the movement of magnets through the stator. This alternating source generated is then converted into DC. This can change batteries which store the electrical energy or can feed into a grid interactive inverter for feeding power into electricity grid. Fig. 9.3 represents the generation of electricity from wind energy.

9.3 Methodology

9.3.1 Software design

The software used for the simulation of this work is LABVIEW [6], which can be used for resolving real time problems. LabVIEW programs

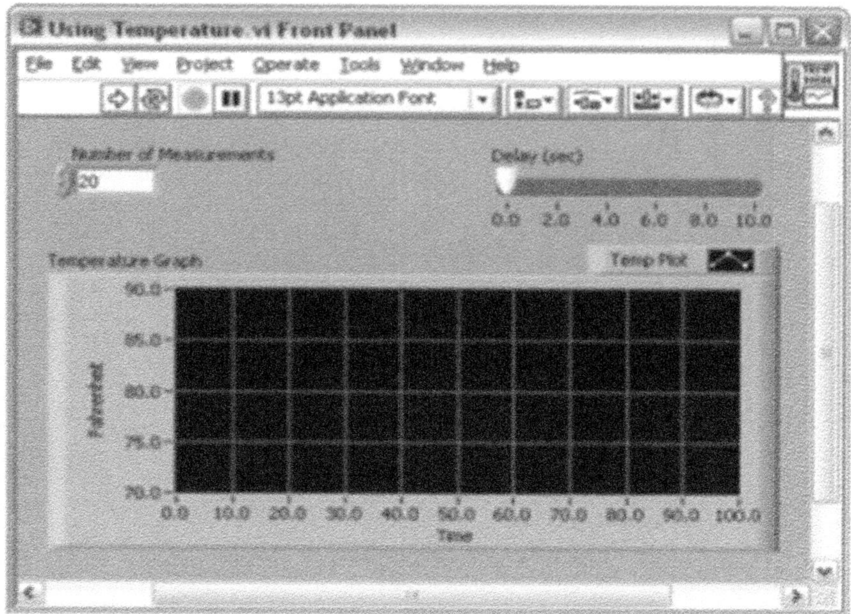

Figure 9.4 Front panel window.

are called virtual instruments, or VIs, because their appearance and operation imitate physical instruments, similar to oscilloscopes and multimeters. LabVIEW contains a comprehensive set of VIs and functions for acquiring, analyzing, displaying, and storing data, as well as tools for troubleshooting the code. LabVIEW VIs contain three main parts—the front panel window, the block diagram, and the icon/connector pane. The window in the front panel is the VI's user interface. A sample of the front panel window is shown in Fig. 9.4.

Once the front panel window is created, code is written to supervise the objects in the control board through graphical representations of functions. An illustration of the front panel window is shown in Fig. 9.5. The graphical source code is available in the block diagram window. The system diagram shows that the front panel objects appear as terminals.

9.3.2 Hardware design

9.3.2.1 Specifications of wind turbine

The wind turbine which has three blades with upwind horizontal- with a rotor diameter of 82.5 m. The rotor has a hub height of 80 m wherein the rotor of turbine and the nacelle were housed on top of the tower [7]. An active yaw and pitch control were employed in the machine with a

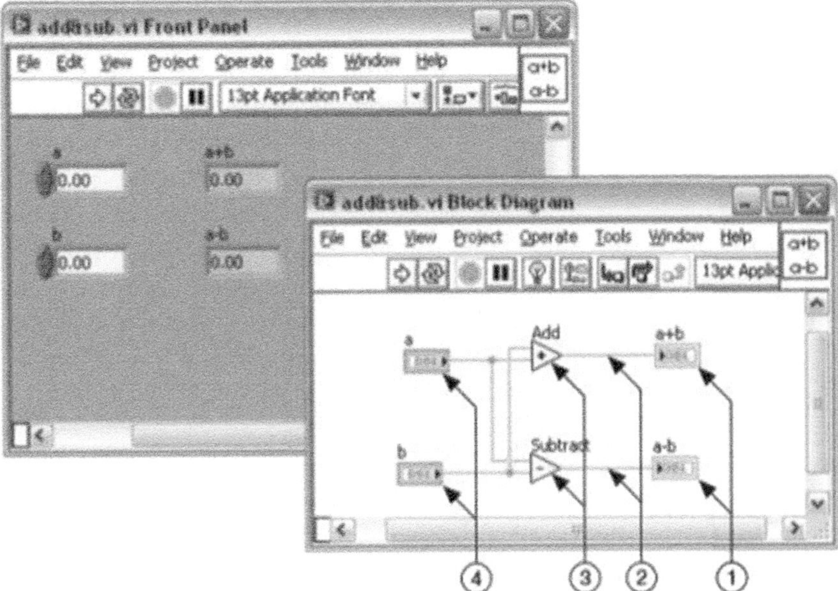

Figure 9.5 Example of a front panel.

generator/power electronic converter system. A distributed drive train is attached to the bedplate of the wind turbine. This drive system includes bearings of the main shaft, gearbox, generator, yaw drives, and control panel. The diameter of the rotor is 82.5 m and having a sweep area of 7389 m and the rotation of operation per minute is designed within 9.8 and 18.7 rotations per minute. The combined action of the pitch angle adjustment of the blade and the torque developed by the generator regulates rotor speed.

When observed from the upwind direction, there will a clockwise rotation of the rotor during normal operating conditions. The range of pitch angle of a complete blade is 90 degrees, and the chord line of air foil flat having 0°-position with reference to the present orientation of wind. The aerodynamic braking of rotor is done when the vanes are pitched at an angle approximately 90 degrees. But the speed of the rotor is limited when the vanes spill. The wind turbine uses three rotor blades. The movement of air foils through the span of the blade having thick air foils provided inside toward the pivot and slowly narrowing down along tip of the blade. During operation, the pitch angle adjustment of the blade will be done by three independent electric pitch motors and controllers. The rotor hub possesses an electric drive which adjusts the pitch angle of the blade which is then joined to a gear ring kept on the internal channel of the pitch bearing of blade. When the speed of wind turbine is above the rated speed, then the active pitch controller triggers the wind turbine to control the speed of wind, by spilling more aerodynamic lift on the blade. When the velocity of wind is less compared to rated one, then speed is captured to make rotor pick up the speed. This energy of the wind from the rotor is converted to the driving force. The machine will stop functioning during fault conditions which may be accomplished using three supporting sections through which the blade's pitch system may be powered. Aerodynamic braking capability will be rendered by providing independent pitch systems for all three blades. The power is transmitted by the gearbox in the wind turbine connecting the rotor and generator. The gear box holds a manifold spiral design. The contained gear train will be housed on the bedplate of the machine. The main braking system will be the individual blade's pitch system of the wind turbine which is electrically actuated. During standard conditions of operation, braking will be applied by pluming the vanes away from gust. A doubly fed induction generator is used here which satisfies the requirements of protection given by the global level IP 54. The bedplate houses the generator and also designed for reducing the noise and vibrations to be transferred to the bedplate. The yawing motion is provided by a roller bearing which is attached connecting the enclosure and steeple. Planetary yaw drives mesh with the outside gear of the yaw bearing and steer the machine to track the wind in yaw. During turbulence, the yaw drives are prevented to attain peak loads from the turbulent wind by engaging the automatic yaw brakes. The sensor of the blade mounted on top of the enclosure gives information to the control unit that activates the drivers of yaw so that the nacelle gets aligned depending on the orientation of gust. Using either an fusing from the enclosure or through a control box placed at the foot of the steeple, an automatic or manual control of the wind turbine can be done. Supervisory Control and Data Acquisition System (SCADA) is used to send control signals from a remote computer with local lockout provided at the turbine controller. The wind turbine uses a power converter system that consists of a converter on the rotor side, a DC intermediate circuit, and a power inverter on the grid side. The converter system consists of a power module and the associated electrical equipment. The generator operates depending on the Variable frequency output from the converter. Table 9.1 shows the technical specifications of the weg.

1. The total output of the wind electric plant is 46.4 MW at site conditions and gross output of each WEG is 1600 kW.

Table 9.1 Technical specifications of the weg.

Parameter	Description
Rotor diameter	82.5 m
Tower height	80 m
Nacelle	Glass fiber reinforced plastic cover with sky light hatch, lighting and venting
Noise abatement	The design of gear box and generator mountings are made so as to minimize the noise and vibration which would be transferred to the blade plate, additionally nacelle housing interior is lined with thick sound insulation material
Lighting	Inside nacelle and tower interiors
Tower	Tubular, steel tower of hub height 80 m designed to IEC specifications. Connection to foundation via to foundation rings. Anticorrosion protection provided by multiple coatings
Door	Steel door including locking cylinder and air louvers
Rotor blades	Three rotor blades with lighting protection system. Rotor diameter 82.5 m. Pitch regulation with electric single blade positioning and supply back up in hub
Gear box	Planetary spur combination
Generator	Doubly fed asynchronous generator with slip rings
Down tower assembly	Integrate the main control, IGBT convertor and low voltage distribution in to single cabinet. Data is designed free of water invasion or tide attack
Power factor of system	The adjustable pf 0.95 inductive to 0.95 capacitive can be maintained at a constant level across the entire power range with in the frame work of control tolerances
Electrical system	As per seller's technical specification
Control cabinet	Design of system controls to protection level IP 54
Wind direction gage and anemometer	Positioned on nacelle roof
Color	Light gray, RAL 7035 or similar, for tower, nacelle and blades
Remote monitoring	Remote data monitoring via modem and computer. The connection is through broadband line. In the event of fault in the operating sequence of the unit, a message is automatically sent to the computer at service center or operating center
Wind SCADA	Wind SCADA Standard over view
Codes and standards	As per applicable IEC Standard and GL guide lines
Alternator output	690 VA
Power factor	0.8
Voltage	33,000 V $\pm$ 5%

2. The protection devices such as that LT & HT switchgear, transformer, switchyard equipment, 430 V MCC, DC system, and lighting, UPS, emergency power supply, and earthling etc., are possessed by the electrical system.

3. The control and instrumentation part of the plant houses interlocks and protection, auto control, local instruments and Programmable logic controllers.

Fig. 9.6 explains about the interconnection of DFIG [8] to grid. To achieve variable speed of the doubly fed induction generator, a stator flux–oriented vector control is used. The magnitude of the emf of stator is altered and made equivalent to the grid voltage amplitude of the by controlling the generator excitation current. The pitch angle control of the turbine triggers the generator till it reaches the synchronous speed to equalize the generator and grid frequency. A large phase difference is produced between two voltages if there is a small difference in frequency of stator and grid is encountered. A phase locked loop (PLL) algorithm is utilized to compensate for the phase difference produced if any. Once the synchronization is achieved, the generator is controlled by connecting to the grid and maximum power will be extracted.

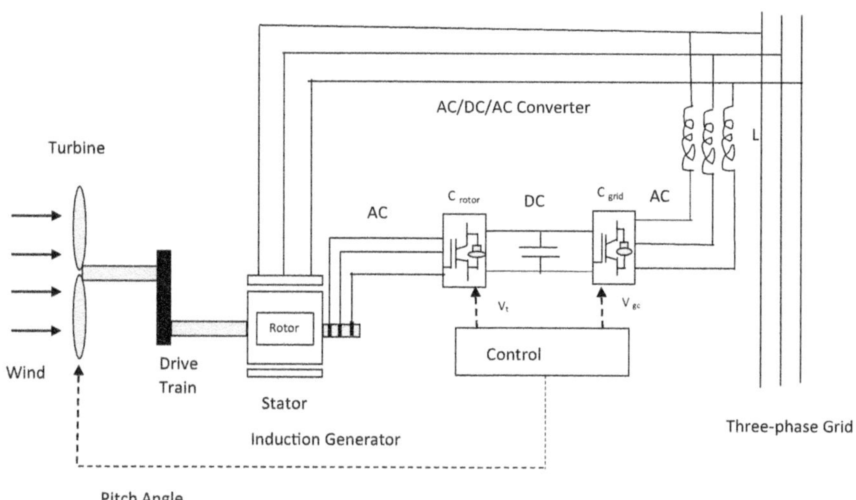

Figure 9.6 Interconnection of DFIG to grid.

9.4 Implementation

9.4.1 Simulation of wind power system

Fig. 9.7 shows the Real time simulation system of wind power system. In the tab control we have two buttons one indicates the speed of Wind and the other, method. The speed of wind is taken as input for various values. The power grid house controllers, blades and drive wherein frequency can be controlled, and output is shown for

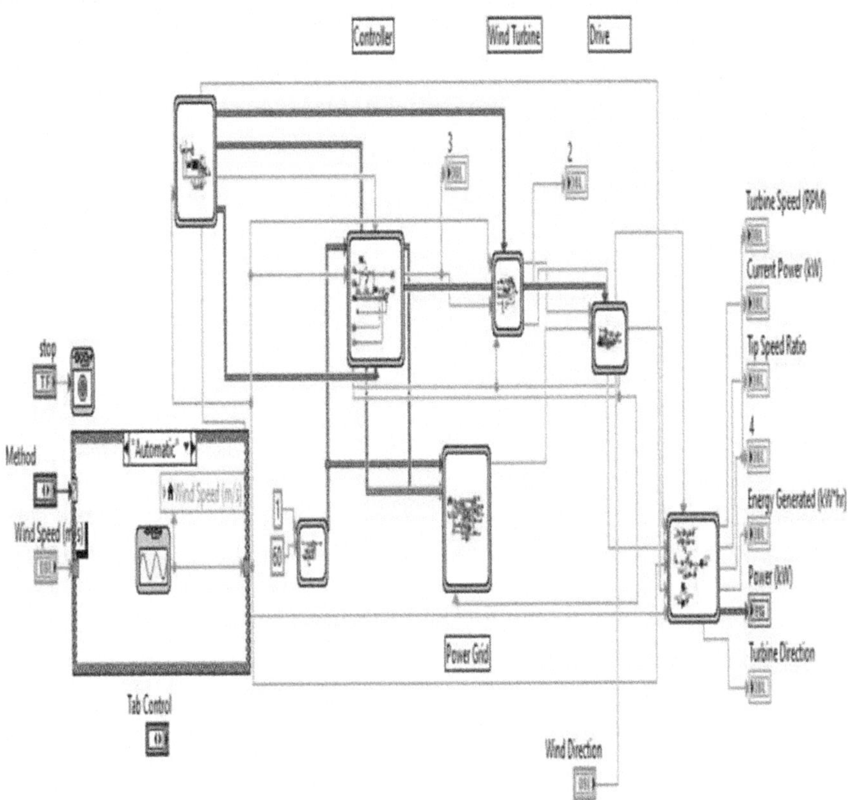

Figure 9.7 Real time simulation system of wind power system.

different values of wind speed. The various outputs for the input wind speed are:

9.4.1.1 Tip speed ratio

The ratio between the tangential speeds at the tip of the blade and the actual wind speed is the tip-speed ratio, λ. The efficiency can be known from the tip-speed ratio, with the optimum varying with blade design. Higher the tip speeds, higher will be the noise levels and this needs a stronger blade due to high centrifugal forces.

9.4.1.2 Turbine speed

The speed of a symmetrical turbine that generates one kilowatt power under one meter is called the speed of a turbine. The formula for specific speed (N_s) of a turbine rotating at a speed of N rpm that develops the power of P kW at the heat of H m is given by,

$$(N_s) = \left(N\sqrt{P}\right) \times (H)\left(5/4\right)$$

9.4.1.3 Power output

Power output is the least amount of wind speed that is generated by the turbine at rated power. It is vital to have knowledge of power and the energy produced by various wind turbines at various conditions.

9.4.1.4 Turbine direction

Wind turbines operate under variable direction and there is an instrument for detecting wind direction. It serves as a return signal for an active system that turns to the nacelle to put the rotor in front of it. At a particular location, there is normally a dominant direction, but also large variations in it, so that this system is standard in current wind turbine designs.

9.4.2 Simulation of substation

The place where part of generation, transmission, and distribution of electrical energy takes place is the substation. The main functions of these substations include transmission of high voltages to low voltages, vice versa. Fig. 9.8 illustrates the simulation of substation in the lab view, which gives us the detailed explanation of each part.

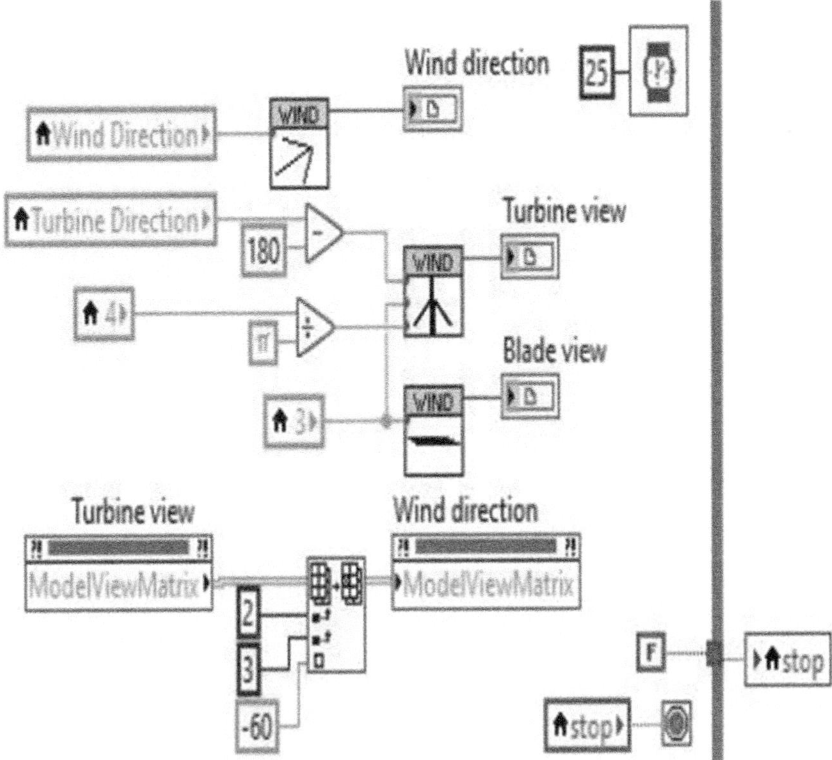

Figure 9.8 Simulation of substation.

Fig. 9.9 shows the experimental set up of the project in LabVIEW

Speed of wind, is a basic movement of air from higher pressure to lower pressure with respect to the changes in temperature. The direction of orientation of the wind gives the exact direction of its flow. Generally, a wind from the northern side moves from north to south. The direction of wind is usually expresses in cardinal directions or in azimuth degrees. The direction of wind is measured clockwise from north in degrees. Fig. 9.10 indicates the speed and orientation of wind in LabVIEW during experimentation.

The turbines used for electric power generation are most often directly coupled to their generators. The speed of generators will be generally maintained at 3000 RPM for 50 Hz systems and 3600 RPM for 60 Hz systems since the generators are required to revolve at

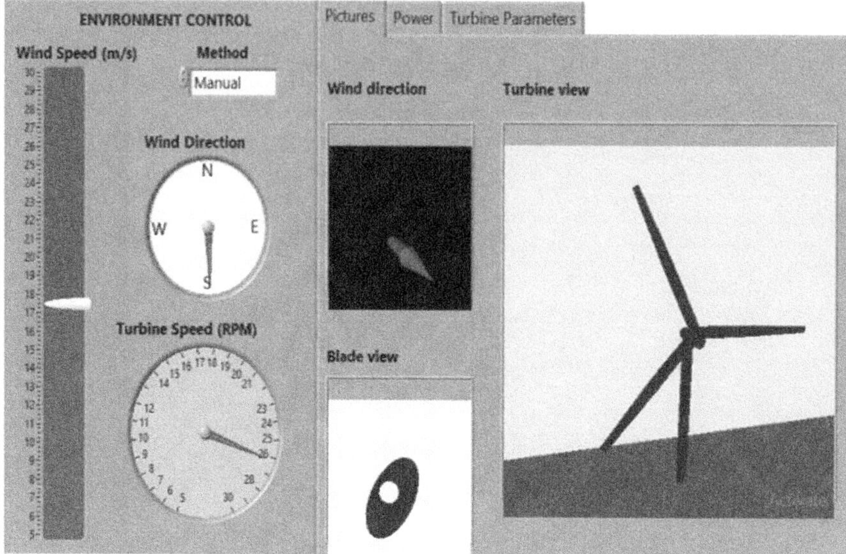

Figure 9.9 Experimental set up in LabVIEW.

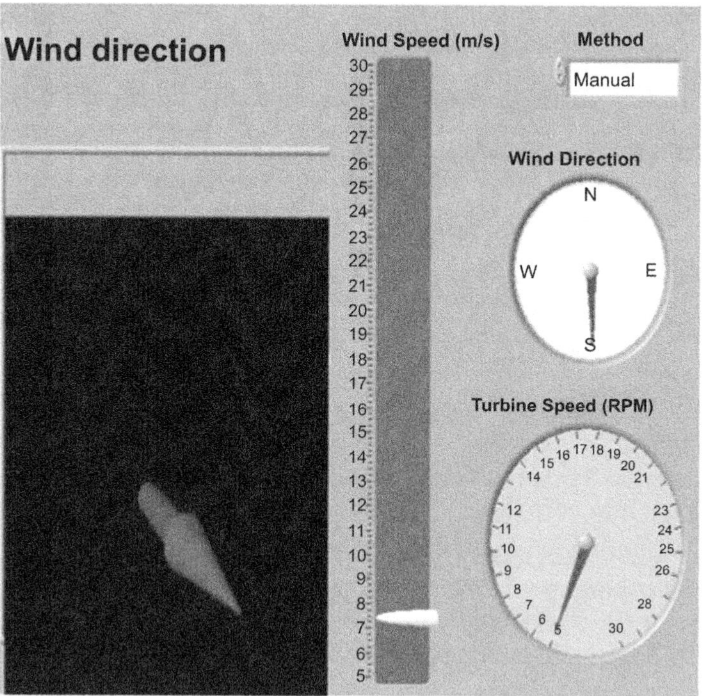

Figure 9.10 Speed and orientation of wind in LabVIEW.

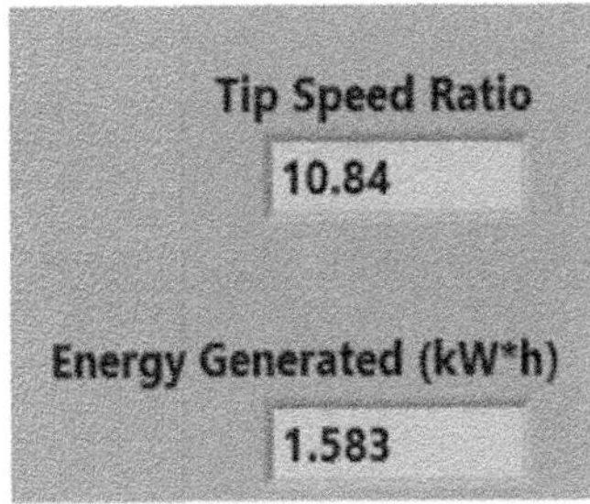

Figure 9.11 Tip speed ratio.

constant concurrent speeds with respect to the frequency of the power system. To design the wind turbine generators, calculation of tip speed ratio is highly essential. Most of the wind may get diverted through the gaps of rotor blades if the wind turbine rotor rotates too slowly or the blades may not move if the rotor rotates too fast. Therefore, it is necessary to design the wind turbines with optimal tip speed ratios so that maximal amount of energy can be drawn from the wind. By dividing the speed taken by the tips of blades of turbine by the wind speed, the tip speed ratio can be obtained. An example says that for a wind speed of 20 mph on to the turbine with the rotation of turbine blades at 80mph, the tip speed rotation can be obtained $80/20 = 4$. Fig. 9.11 shows the tip speed ratio.

9.5 Simulation results

Fig. 9.12 shows the outputs of the simulation. With the speed of wind and the pitching angle taken as the inputs, the power and torque can be calculated. The coefficient of power will be calculated in terms of tip speed ratio. Through the VI interface, it is possible to alter the load. This provides an easy way to realize the dynamics of a wind turbine. The response shows that performance of the system through the speed of rotor, penetration angle and coefficient of power.

The production of power by the WT varies from hour to hour and minute to minute even though the site selected for deployment is the best. Depending on the demand from the user, the grid should retaliate. Wind turbines on the grid do not contribute to meeting demand since the grid dispatchers cannot control the production of

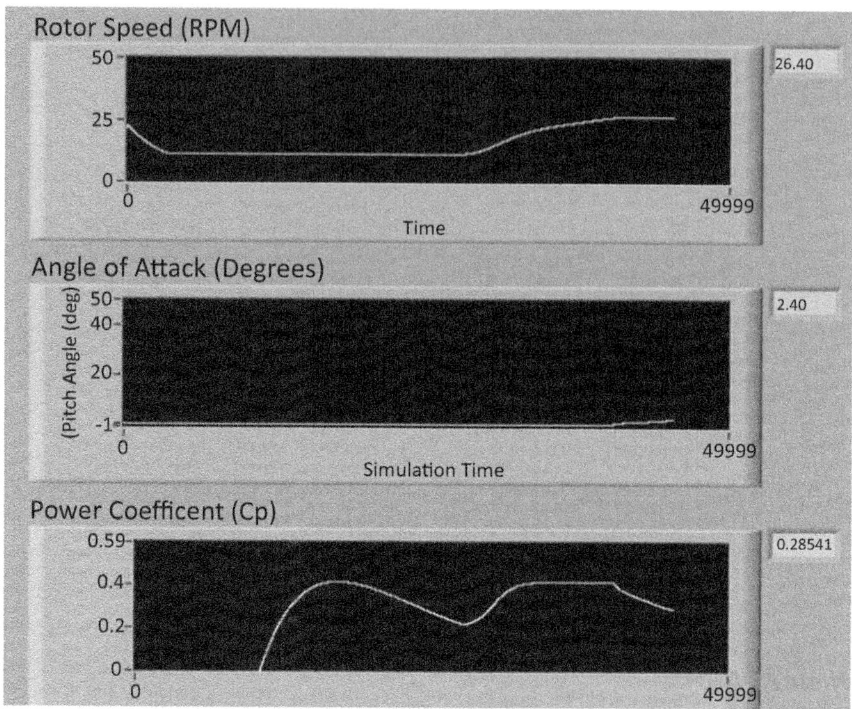

Figure 9.12 Output of rotor speed and power coefficient

wind power than controlling the user demand. So by allowing power to the grid, a source of fluctuation will be included which says the grid to balance. The General Electric (GE) fabricated a 1.5-MW model which was used widely once. This 1.5 MW is its rated, or maximum, capacity, at which rate the power is produced when there is an ideal range of wind between 27 and 56 mph. The general range of turbines at present is in the range of 2−3 MW. The windmill gives the power by extracting the energy in the, so the capability of capturing the energy and converting that to a torque gives the power of turbine. This can trigger the generator and provides electricity to the grid. Huge is the size of the blades, high will be the capture of wind energy. Taller the tower, steadier will be wind captured. Depending on the size of the generator, the blade size changes and will be the capture of wind. Fig. 9.13 shows the output power obtained during simulation using LabVIEW.

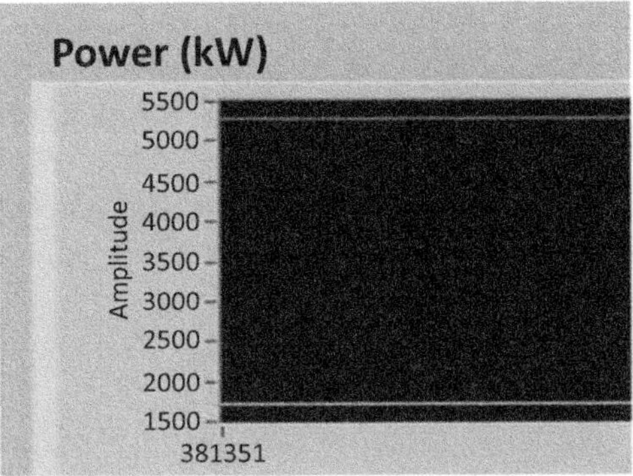

Figure 9.13 Output power.

9.6 Conclusion

Wind electric systems are complicated and require several integrant arranged in subsystems having different control strategies wherein modeling and performance analysis necessitates broad range of timescales. In this project the characteristics of wind turbine are explained by simulating in Lab VIEW. The speed of wind and the inclination angle are taken as inputs to estimate the power and torque produced. Depending on the tip speed ratio, the coefficient of power will be evaluated constantly. The graphical user interface is deployed to alter the load and to have easy realization of turbine dynamics. The result of simulation showcases the efficient performance of the windmill. A feasibility study was accomplished on the power generation of wind hybrid generation system connected to grid wherein the proposed study exhibits features such as power electronic interfaces and control schemes for maximum power generation. The models for analyzing the dynamics of the present system with power control schemes were also addressed and simulations were carried out. The results of simulation show the robust performance of wind control system even under acute variations in wind speed conditions.

9.7 Future scope

The future scope of this work is elaborated in detail:
1. Hybrid models for wind farm
   A combination of physical and statistical modeling is included in present-time researches to forecast the wind power [9].

2. Models for wind power ramp forecasting

   With the high perforation of wind power, the grid may encounter the challenge due to the variations in wind power. The ramp event is a very critical issue, and it is characterized by a sudden large change (increment or decrement) in wind power. To maintain stability of the grid, it is very vital to have perfect models for detecting the ramp event.

3. Estimation of forecast uncertainty

   The integration of grid and functioning of the system requires the analysis of uncertainties in the forecasting.

4. Wind energy commerce under uncertainty in electricity markets

   Perfect forecasting of wind power leads a wind farm to a wind power plant. This facilitates the operators of the wind farm to market electricity with good profit compared to the use of traditional sources.

5. Condition monitoring of wind farms

   To estimate the robustness of the wind turbine, better forecasting models for the wind power and power curves for the wind turbine may be used.

6. Wind forecasting and offshore

   Many countries in the world already initiated the offshore wind farms due to restricted land areas. These offshore wind farms have opened up huge prospects of research in operation, maintenance, control of wind farms, and wind resource assessment to identify potential sites, etc.

Conflict of interest

There are no conflicts of interest to disclose.

References

[1] G. Kaur, D. Pardeshi, B.S. Surjan, Literature review on wind power generating system, International Journal of Innovative Research in Electrical Electronics Instrumentation and Control Engineering 4 (2016) 6. June.

[2] R. Thresher M. Robinson and National Renewable Energy Laboratory P. Veers Sandia National Laboratories, Wind Energy Technology, in: Proceedings of the Physics of Sustainable Energy Conference University of California at Berkeley March 1−2, 2008.

[3] V. Aiswarya et al. Wind turbine instrumentation system using Lab VIEW, in: Proceedings of IEEE Global Humanitarian Technology Conference: South Asia Satellite (GHTC-SAS), 2013.

[4] S. Vaishali, et al., A critical review on wind turbine power curve modeling techniques and their applications in wind based energy systems, Journal of Energy 2016 (2016). Article ID 8519785.

[5] G. Kaur, et al., Literature review on wind power generating system, International Journal of Innovative Research in Electrical, Electronics, Instrumentation and Control Engineering 4 (6) (2016) 246−249. 2321 − 2004.
[6] <https://www.ni.com/getting-started/labview-basics/environment>.
[7] <http://forestsclearance.nic.in>.
[8] A.G. Abo-Khalil et al. Grid connection of doubly-fed induction generators in wind energy conversion system, in: Proceedings of IEEE Fifth International Power Electronics and Motion Control Conference.
[9] Md. A. Hossain, et al., Very short-term forecasting of wind power generation using hybrid deep learning model, Journal of Cleaner Production 296 (2021).

CHAPTER 10

Cyber attacks, security data detection, and critical loads in the power systems

S. Rama Devi, P.S. Latha Kalyampudi and N. Sai Charitha
BVRIT HYDERABAD College of Engineering for Women, Hyderabad, Telangana, India

10.1 Introduction

All major sectors such as construction, transport agriculture, extraction of natural resources, manufacturing, health services, and communication are dependent on a reliable supply of electricity in a macroeconomic system. Electric power supply is essential in the economic development of the entire world (Fig. 10.1), which shows the smart grid architecture, where there is a central control that can control the generation of power through numerous sources such as wind, coal, solar, nuclear and others. This power is then sent to various distribution centers and then to consumers, such as households, universities, businesses, and others. To improve efficiency of the smart grid, there are five important factors to be considered,

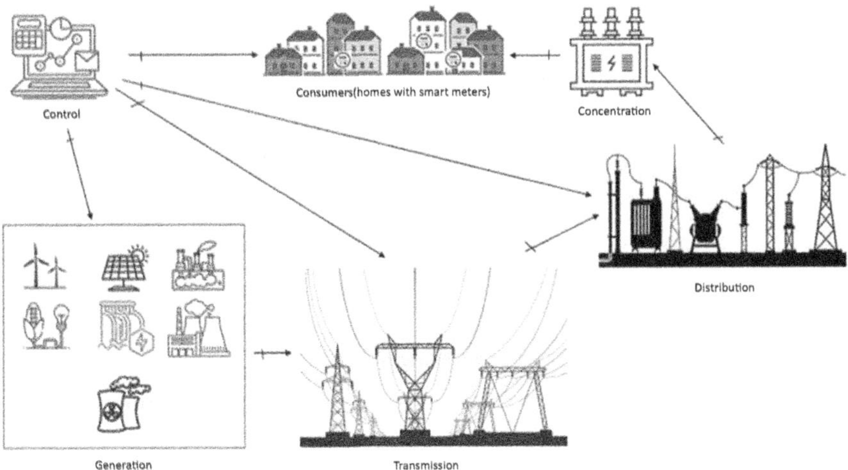

Figure 10.1 Smart grid layout.

Smart Energy and Electric Power Systems
DOI: https://doi.org/10.1016/B978-0-323-91664-6.00012-7

such as smart metering, communications, monitoring, distributed energy resources, and controlling [1].

The economic and social losses may occur due to targeted attacks on power networks and which will disturb the operations of the smart grid. Owing to cyber attacks on the smart grid, time-critical data can be lost and modified, which will have an impact on real-time operations and decision-making. For the sustainable operation of power systems, many protection and control mechanisms are required. Power system controllers should be able to ensure power system stability. Next chapter mainly focuses on a list of cyber attacks and their effects on the reliable functioning of the electric grid. This also helps system operators understand the data attacks and take effective countermeasures.

For the proper functioning of the grid, several main security objectives such as confidentiality, Availability, and integrity must be included [1]. According to user requirements Availability of an uninterrupted power supply, it ensures reliable and timely access to information. The integrity of information means protection against illegal and improper modification of data and confidentiality prevents unauthorized access of data. Types of attacks, primary weaknesses, diverse attackers, and required security solutions must all be addressed. The exploration of the behavior of the power grid under disturbances is known as the power system stability. After being subjected to some sort of disturbance, the tendency to return to a stable operation. Steady-state, transient, and dynamic stability are the three types of stability:

1. The behavior of an electrical machine to a constantly growing load is called as steady-state stability.
2. The response that happens as a result of minor system changes, leading in oscillations is called as dynamic stability.
3. The reaction to large disturbances that can result in significant changes in rotation speed, energy transfers, and current angles is called transient stability.

The three crucial quantities that should be effectively controlled to ensure the power system's stability are frequency, rotor angle, and voltage.

10.2 Cyber attacks in smart grids

10.2.1 Phishing

Phishing is a kind of social engineering attack by collecting personal information using websites and emails as sources. Phishing is used to steal credit card numbers, user data, including login credentials. Phishing is a method of cyber attack that employs the deployment of a spoof email as a technique. The idea is to persuade the email recipient that the message contains something they desire or need to open the link or open an attachment, such as a request from their financial institution or a message from a friend. Phishing and zero-day exploit attacks allow attackers to enter into a system try to steal sensitive information, which will result in damage to the system [2].

Phishing is a simple process to carry out, because it is the first stage in putting individuals and clients at risk. The attacker tries to act like a trusted entity and makes the victim open email and text messages. The recipient tries to open the link sent by the attacker which results in the installation of a virus or revealing confidential information. This confidential information about the power supply will be used by the attacker and inside the organization, Personal information may be entered by the employee via fraudulent emails or messages that appear to be legitimate emails sent by the hacker, which can lead to hacking. These hazards are likely to harm a smart grid client, as giving information to unknown sources and without knowing the implications of these risks might have a mental and financial impact on customers. However, it is a significant worry when dealing with anti-phishing security measures [3].

10.2.2 Denial-of-service

Any attack against availability security feature comes under the denial-of-service (DoS) attack. Using a distributed network architecture, the smart grid delivers power to numerous devices across a large area. The connectivity must be secure and reliable otherwise loss will be immense. This will result in huge losses leading to catastrophe if there is a failure in the grid. The collapse of a grid can result in massive losses and even massive failure. The hacker tries to jam the channel by modifying the MAC address and using tools to gain access into the network in this DOS attack [4]. The network will be flooded with so many packet requests to the computers. That means the hacker could be almost anybody, from a single person to a larger group of people.

Security protocols available in the network and transport layers, such as transmission control protocol (TCP), IPV6, and secure sockets layer (SSL), are also vulnerable in the smart grid network design [5]. DOS attacks can interrupt electrical systems by overwhelming computer systems with information in a bid to take them down. DOS attack uses one source for attack whereas distributed DoS or DDoS uses multiple hosts as sources for delivering a heavy stream of information into the network which increases the attack and makes defense very difficult.

Flooding attempts and security vulnerability attacks are the two forms of DDOS attacks. SYN flooding and Internet control message protocol flooding are two types of flooding attacks. Sync flooding takes the advantage of the TCP for establishing the connection between the hosts [5]. To make a connection, the host sends a request. The attacker sends SYn. Not to the actual host, but packets with source addresses. The target responds by sending an ACk packet to the source address. The target continues to wait for a packet acknowledgment to finish the connection process because the source address was invalid. This results in DDO Attacks because of the allocation of resources for these malicious packets [6].

10.2.3 Malware spreading

The other kind of attack in the smart grid is by malware spreading [7]. The attacker can develop software or firmware and inject that into company servers and devices for getting confidential data of smart meters. This kind of unauthorized process will have significant impact on integrity, confidentiality, and availability of information. Trojan horses, viruses, worms, and spyware are examples of malicious code.

10.2.4 Eavesdropping and traffic analysis

Spoofing attacks include eavesdropping and traffic analysis. The attacker can gain access to sensitive information by monitoring network traffic. The smart grid is vulnerable to this attack as it is made up of a huge network [8].

10.3 Smart grid cyber security needs and standards

Cyber security is an important measure to be considered for the deployment of smart grid. In the smart grid deployment, the cyber security standard adoption should be an integrated element of the blueprint and development process. The adopted cyber security strategy should guarantee that legacy systems are well protected. A best cyber security policy will result in smooth functioning of grid environment without loss of critical data while yet allowing for the operational efficiency and flexibility.

Smart grid must create a thorough, integrated, monitored, and regularly updated cyber security system to ensure that they get secure processing [9]. Cyber security is needed for different core components of the smart grid's operational systems such as SCADA systems, operational control systems, substations and smart meters.

10.3.1 Risk assessment

A comprehensive operational risk assessment is a beginning stage of cyber security program. The risk assessment process in the smart grid is based on current risk assessment strategies being implemented by both the government and industry, and it entails identifying various kinds of assets, security flaws, and threats, as well as specifying impacts, to evaluate a risk rating to the smart grid and its important domains and subdomains, such as homes and businesses [10]. The risk assessment procedure is done to all sectors in the smart grid as they interact. From an operational standard point of view, the Analysis enables an Operator to identify possible problem areas and then build a robust cyber security policy for smart grid technology adoption.

The below measures were used in the risk assessment process in developing of the smart grid:

1. NIST Special Publication (SP) 800-39, Managing Information Security Risk: Organization, Mission, and Information System View, NIST, March 2011;
2. SP 800-30, Risk Management Guide for Information Technology Systems, NIST, July 2002;
3. Federal Information Processing Standard (FIPS) 200, Minimum Security Requirements for Federal Information and Information Systems, NIST, March 2006;
4. FIPS 199, Standards for Security Categorization of Federal Information and Information Systems, NIST, February 2004;
5. Security Guidelines for the Electricity Sector: Vulnerability and Risk Assessment, version 1.0, NERC, June 14, 2002;
6. Systems Part 1: Terminology, Concepts, and Models, International Society of Automation (ISA), 200711; and
7. ANSI/ISA-62443-2-1 (99.02.01)-2009, Security for Industrial Automation and Control Systems: Establishing an Industrial Automation and Control Systems Security Program, ISA, January 200912.

The Smart Risk Management Methodology framework is a seven steps based, the framework outlined in ISO/IEC 27005 shown in Fig. 10.2. The initial phase in the Operational Risk Assessment is to undertake an impact assessment, followed by a Threat & Vulnerability Risk Assessment [11] to define priority and target areas for protection, risk calculation and prioritization. An infrastructure review, an examination of security systems, network equipment, servers, and terminals, a cyber security process evaluation, and a plant audit, including a physical vulnerability scanning are all part of the Operational Risk Assessment.

In the risk assessment can also include its information systems for overall productivity, protected from cyber-attacks. The cyber security standards and industry-specific security requirements should be considered for operational risk assessment. A thorough cyber security program should be designed when the appropriate risk assessment processes have been completed. The chosen implementations must be based on open standards, to ensure that the smart grid is safeguarded by effective and cost-effective technologies that complement and collaborate in a highly efficient manner. Interoperability with legacy operational systems is also a need for the chosen solutions. The cyber security program will be able to develop,

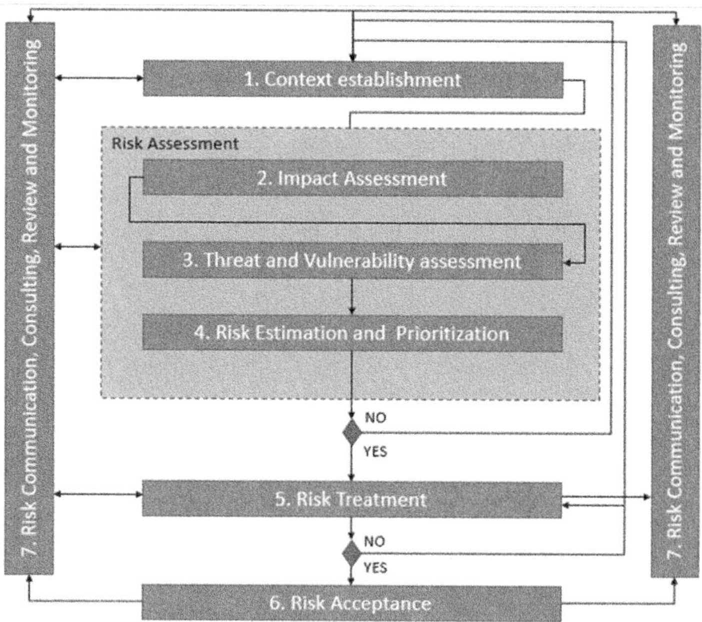

Figure 10.2 Security for Smart Risk Management Methodology.

expand, and adapt to new difficulties as they occur thanks to the use of open-source technology [12].

10.4 Proposed security solutions for smart grids

The security risks require security solutions to protect against vulnerabilities, the smart grid network security; networks are the most vulnerable to threats and risks [13]. Threats and risks are the biggest barriers to network and system maintenance. Because all networks are based on the OSI-Model, an attacker can use other network layers of the model (Fig. 10.3). Appropriate security solutions should be taken to overcome cyber attacks in the grid technology [14].

10.4.1 Encryption

The technique of encoding information is known as encryption. This procedure converts plaintext, or the original data representation, into ciphertext, or an alternate data representation. Only authorized persons should be able to decipher a ciphertext back to plain text and access the original data. The network uses a VPN to connect to the internet, your connection is encrypted, when an attackers try to intercept the data stream would only get nonsensical, the encrypted data using a key (or logic) only the authorized user/receiver can be decrypt the message. The encryption can be Symmetric and asymmetric. If the same key is used to encrypt and decrypt data in symmetric encryption. The encrypted message in communication has two aspects: the sender encrypts the data and the receiver decrypts the data with the key. In the asymmetric encryption the sender and receiver use different keys. This technique is also known as public-key cryptography, uses a public key-private key pair, where data

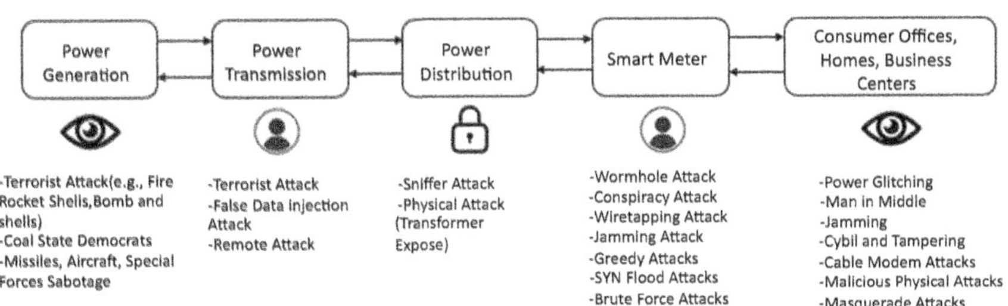

Figure 10.3 Types of attacks on different layers of smart grid.

encrypted with the private key can only be decrypted with the public key and vice versa.

SSL or transport layer security (TLS), a technology that allows the use of HTTPS, uses asymmetric encryption [15]. A website's public key is obtained by the client from a TLS (or SSL) certificate for that website and used to establish a secure connection. Your private key is hidden from the website. As a result, algorithms are used to encrypt or decode data. Different algorithms can be used depending on the level of security required. Some of the most commonly used symmetric algorithms are AES, DES, and SNOW, where as asymmetric algorithm RSA, Elliptic Curve Encryption are the most commonly used techniques.

Advanced encryption standard (AES) (Fig. 10.4) is one of the trusted standard algorithm by the US government and many other organizations and also mostly recommended VPN providers [15]. AES encryption is very effective for 128-bit formats but is also suitable for 192-bit and 256-bit encryption for stronger encryption. AES is said to be resistant to all attacks, except for brute-force attacks that attempt to decrypt messages using any possible combination of 128-bit, 192-bit, or 256-bit ciphers.

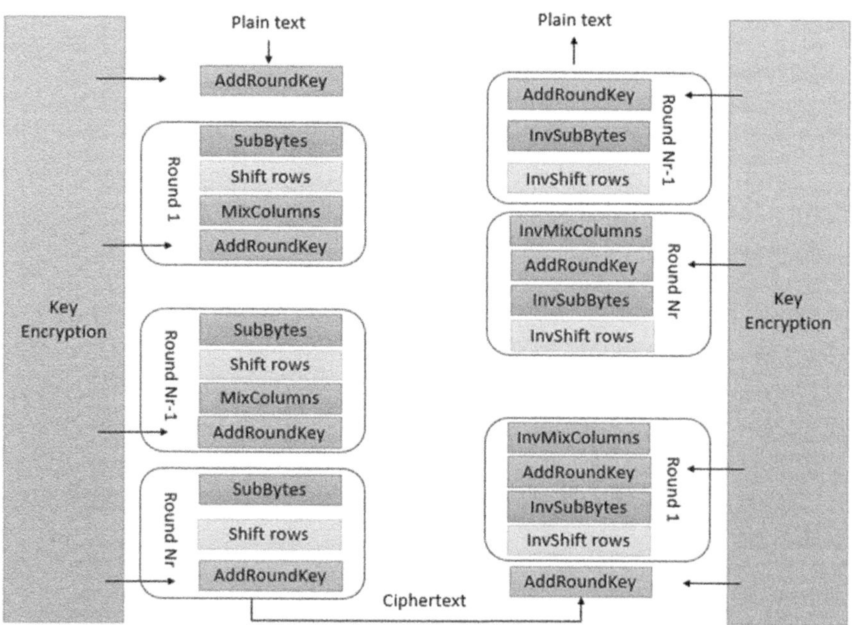

Figure 10.4 The overall structure of encryption and decryption in advanced encryption standard.

The 256-bit refers to the size of the encryption cipher. The greater the size, the greater the number of possibilities, and the more difficult it is to guess the key. The number of possible 256-bit encryption combinations is greater than the number of stars in the universe. This kind of encryption is needed to secure banks and governments all over the world to protect their sensitive data. The AES algorithm implemented on the ESP32 with the LoRa unit, and validation was obtained by generating a single Message Authentication Code (MAC) to secure wireless communication for micro-grids. All 4G networks currently use 128-bit AES encryption, and future 5G networks will use 256-bit AES encryption.

The Public Key Infrastructure (PKI)-based system automates encryption and decryption methods to assure data confidentiality and integrity between smart meters, gateways, and back-ends in smart grid [16]. With the help of these measures, when implemented at the core of edge devices, can prevent data interception and tampering along the route, which could affect system validity or even put the grid at risk. The Cryptographic countermeasures for the smart grid also include key management at various scales, ranging from tens (substation) to millions of credentials and keys (AMI). Improper information security compromises the complete goal of secure communications in the smart grid.

10.4.2 Authentication

Authentication is a process that identifies a user or a computer system using some factors that the user decides on before the identification. Fingerprints or birthmarks are simple examples of authentication. Fingerprints and birthmarks are unique to each individual and can be used to identify them. Similarly, for computer systems, there is a login page where you enter a unique user ID and an associated password that allows identifying the user. According to the authors, the main concern is maintaining authentication and control access, with the strong authentication mechanisms the identity should be verified. Authentication is a critical process for preventing data integrity attacks.

Smart grid authentication can be accomplished in a variety of ways. These methods are classified as cryptographic, password, hardware, signature, and biometric. The cryptographic approach is thought to be the most secure for smart grid authentication. The key requirement for authentication in smart grid applications is the implementation of robust multicast authentication mechanisms. Public key-based multicast authentication [16] is a modern security standard for substation communication that is also suggested by IEC 62351. In public key-based multicast authentication, all receivers share the sender's public key.

The authentication can be given by adopting standard policies like "implicit deny policy" for network access where grant access only to explicit users. Using policies that provide a security solution for your organization and using implicit deny policies can be useful because individual users have different permissions that grant specific permissions to individual users. Here the manager can see all the additional data related to the project, but the staff is limited. access to data. The risk of breaches can reduce by giving explicit access to individuals in organization to identify who has access to your network. Smart grid networks require higher bandwidth for communication, which means they can also use encryption methods for authentication. Maintaining an encryption method increases costs but provides a good authentication mechanism [17]. It is best for management systems and IT security engineers to work together to secure the smart grid.

10.4.3 Malware protection

Malware is malicious software (Viruses, worms, Trojan viruses, spyware, adware, and ransomware) that is designed to harm and destroy computer systems. Malware protection is required on all Embedded structures, in addition to General-cause structures. Embedded systems are only intended to run software that has been provided by the manufacturer. The manufacturer is required to include a reliable storage device in its products, as well as keying cloth for software program validation. Before walking it, the device can validate the newly downloaded software program by using a key. The device necessitates the use of up-to-date and frequently updated antivirus software, as well as host-primarily based total intrusion prevention.

The embedded devices and general-purpose systems connected to the smart grid must be safe and secure from cyber-attacks, antimalware protection is required [18]. A manufacturer key is required for embedded systems, which can be used to protect the product during software certification. The main reason embedded systems are more secure than general-purpose systems is that general-purpose systems can run constantly updated third-party software like antivirus software, whereas embedded systems can only run software provided by the manufacturer and require a manufacturer key to verify the software. When providing solutions, manufacturers' software must be genuine, and organizations may need to adopt risk management to mitigate risks when using third-party software. The manufacturer key is also used to verify whether the software is genuine also checks copyright permission. The use of antimalware antivirus software may include some level of protection and basic risk detection techniques. The antivirus software, must be kept up to date with new patches, and attackers must avoid looking for ways to get around antivirus software because threats can be made vulnerable to malware through phishing attempts [14]. The use of a Network Intrusion Prevention System (NIPS) to limit network access and protect it from various cyber assaults is an effective strategy suitable for usage in a smart grid and also used to monitor intrusion data, take action to prevent attacks from progressing. A Network Intrusion Detection System (NIDS), which monitors and analyses network traffic to safeguard systems from network attacks, is another smart grid solution.

10.4.4 Network security

The security risks in the smart grid can be addressed by countermeasures and protection methods which are widely deployed and integrated into network protocols and structures. In power grids, DNP3 and IEC 61850 are two extensively used communication protocols. In the United States, DNP3 is commonly utilized for intra-substation and inter-substation communication. IEC 61850 with secure DNP3 and IEC 62351 is provided to provide secure end-to-end communication in power networks. In addition to these end-to-end security protocols, a secure data aggregation protocol for smart grid is also provided because bottom-up traffic models (device-to-center) such as meter reading and device monitoring of AMI networks are prevalent in power systems. In SCADA networks [19]. In comparison to secure DNP3, IEC 61850 with IEC 62351 is a modern power communication protocol that balances security and temporal importance by using two separate security levels for different sorts of communications in the power system. More complex layered security mechanisms are likely to be proposed to meet both security and quality of service requirements for delivering messages to the smart grid.

In this communication architecture, data aggregation methods inside the network (Fig. 10.5) will be more efficient than end-to-end routing protocols in which each node seeks to discover its own route to the

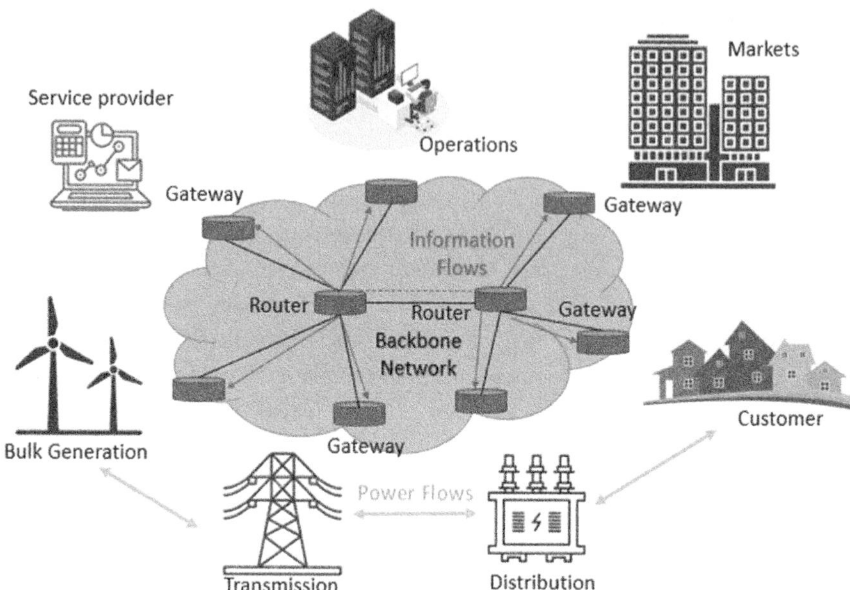

Figure 10.5 The network architecture in the smart grid: backbone and local-area networks.

center. Because secure data aggregation costs more CPU resources and adds additional delay overhead, current research focuses on secure data aggregation protocols for the AMI network, where communication traffic is less time–critical [17].

When using public networks such as the Internet, virtual private networks (VPNs) can provide additional security. This level of network security aids in preventing unauthorized network access. Create a virtual network tunnel so that data can safely travel from point A to point B. VPNs employ use security measures, such as encryption and data protection for all data transmitted over the network, because data can be compromised if: It uses public network infrastructure. There are two kinds of VPNs that are useful for both institutional and general users.

The Remote access VPN and Site-to-Site VPN are two types (VPNs) useful for organizations and end–users.

10.4.5 Remote access virtual private network

A remote VPN connects to a private network via a public network, such as the Internet. The Internet is used to connect users to their private network, which is both safe and private. Both home and business users can benefit from a remote access VPN. After authentication, users gain access to a VPN gateway via a mobile device or desktop computer. If the credentials are correct, access is validated by authentication, and you can then access the virtual network's resources.

These resources, which include business applications and documents, are only accessible to users within your organization. Users can access their work via remote access VPN from any location as long as they are connected to a VPN gateway. Home VPN users primarily use VPN services to circumvent georestrictions on the internet and access blocked websites. VPN services are typically used by home users to bypass geo-restrictions on the internet and gain access to forbidden websites. VPN services are used by those who are concerned about their internet security and privacy.

10.4.6 Site-to-site virtual private network

A site-to-site (router-to-router) VPN is commonly used in large corporations. Organizations having several branch offices in different places utilize site-to-site VPN to connect one office's network to another office's network, allowing safe access or resource sharing to many branches in different locations for the organization's partner or client company. Site-to-site VPN is further subdivided into as Intranet-based VPN and Extranet-based VPN, The Intranet-based VPN that connects various offices of the same corporation. Extranet-based VPN, in which organizations interact to another business office via Site-to-Site VPN. Site-to-site VPNs, in general, construct an artificial bridge among networks at located in different geographical offices and connect them through the Online platform, allowing for safe and private communication. Since Site-to-Site VPN depends on router-to-router interaction, one router serves as a VPN Client while the other serves as a VPN Server. The communication between the two routers begins once the authentication between them is confirmed. The Internet Protocol security (IPSec), Layer 2 Tunneling Protocol (L2TP), Point–to–Point Tunneling Protocol (PPTP), SSL and TSL, Open VPN & Secure Shell (SSH) are the protocols that VPNs rely on in the same way as network protocols do [20].

10.4.7 Intrusion detection systems and intrusion prevention system

A network intrusion prevention system (IPS) detects and blocks known threats. An IPS continuously monitors the network for possible harmful occurrences and gathers information about them. The IPS alerts systems admins from these incidents and takes precautionary measures, such as blocking access points and installing firewalls, to avoid future assaults. IPS systems also can be used to discover vulnerabilities in security policies, as well as to deter individuals and network guests from violating the policies' restrictions. Intrusion detection systems (IDS) are encountered by IPS solutions [20]. The action performed when a possible incident is discovered is the main difference between IPS and IDS.

IPSs control network access and protect it from un-authorized access and attack. IDS systems operate in three ways. (1) The simplest way is signature-based detection, which checks components of activity such as logs or data streams using string operators. (2) To discover large deviations, anomaly detection compares parameters of typical behavior to observed events. This type of detection has the ability to be very helpful in detecting hazards that were previously undiscovered. (3) Stateful protocols analysis compares established profiles of commonly recognized standards for acceptable protocol behavior for each protocol stage to observable events to discover deviations.

These systems are intended to monitor intrusion data and take action to prevent an attack from occurring. IDSs are intended to keep an eye on your network and notify system administrators when possible threats are spotted. Reinforcement learning, Semi-supervised learning, Active learning, and DL solutions can all help IDS. The terminology used to define knowledge discovery, according to data mining, can be utilized as a foundation for designing and implementing IDSs with more accuracy and robust behavior than conventional IDSs that are based solely on specific rules. The concept of an intelligent security mechanism is to create a model that either learns the normal behavior of a system or generates a template for each anomaly, such as a DoS attack or a man-in-the-middle (MITM) attack, etc. The detection engine can identify assaults in real time using this knowledge. The disadvantage of employing machine learning algorithms for classification is that an attacker may attempt to overcome the classifier, resulting in misinterpretation and an attacker's attack. Antivirus and IPSs provide proactive defense against today's most common network assaults. When used correctly, IPS can protect against malicious or undesired packets as well as brute force attacks. For any network, a Next-Generation firewall provides enhanced intrusion prevention and detection [21]. Depending on your needs, you can use an IDS, an IPS, or both. Using either approach data transit can make more secure.

10.4.8 Maturity assessments

Several smart solutions have recently been developed for conducting fast cyber security hazard identification and mitigation strategies in large complex systems and locations where full security inspections are not practicable owing to time and capacity restrictions. Risk assessments can be a continuous process that aids in the protection, discovery, and correction of potential risks or vulnerabilities. Many organizations are proposing a cyber defense triage process to help identify priority areas where the impact of attacks is greatest, in immediate effect appropriate security solutions is applied to the application/areas [10]. Risk assessment is just one of the many controls that can be evaluated when performing a maturity assessment, as proposed, to identify and mitigate security gaps and weaknesses [11].

The security of a secure smart grid must evolve and be updated on a regular basis, with credentials and computerized key exchange challenges revised every 2–3 years. The cyber security program must be updated on a regular basis because cyber security risks are always emerging. To guarantee that effective cyber safety measures are taken always, the operator should commit to constant cyber monitoring and frequent re-evaluations of its environment.

Smart grid cyber security standards have been published by organizations such as the NIST (National Institute of Standards and Technology) and ENISA (European Network and Information Security Agency), which should be considered when establishing a holistic cyber strategy. A smart grid cyber plan should contain mitigation, diagnosis, response, and recovery mechanisms to respond to present and anticipated threats, as per NISTIR report 7628.

References

[1] O. Otuoze, M.W. Mustafa, R.M. Larik, Smart grids security challenges: classification by sources of threats, Journal of Electrical Systems and Information Technology (2018).

[2] H. He, J. Yan, Cyber-physical attacks and defences in the smart grid: a survey, IET Cyber-Physical Systems: Theory & Applications 1 (1) (2016) 13–27. Available from: https://doi.org/10.1049/iet-cps.2016.0019.

[3] S. Tan, D. De, W.Z. Song, J. Yang, S.K. Das, Survey of security advances in smart grid: a data driven approach, IEEE Communications Surveys & Tutorials 19 (1) (2017) 397–422. Available from: https://doi.org/10.1109/COMST.2016.2616442.

[4] L. Kotut, L.A. Wahsheh, Survey of cyber security challenges and solutions in smart grids, in: Proceedings of the Cybersecurity Symposium (CYBERSEC), pp. 32–37, Apr. 2016.

[5] S. Ahmed, T.M. Gondal, M. Adil, S.A. Malik, R. Qureshi, A survey on communication technologies in smart grid, in: Proceedings of the IEEE PES GTD Grand International Conference and Exposition Asia (GTD Asia), pp. 7–12, Mar. 2019.

[6] F. Jameel, Network security challenges in smart grid, in: Proceedings of the Nineteenth International Multi-Topic Conference (INMIC), pp. 1−7, Dec. 2016.

[7] P. Eder-Neuhauser, T. Zseby, J. Fabini, G. Vormayr, Cyber attack models for smart grid environments, Elsevier Journal on Sustainable Energy, Grids and Networks 12 (2017) 10−29. Available from: https://doi.org/10.1016/j.segan.2017.08.002. December.

[8] M.Z. Gunduz, R. Das, Analysis of cyber-attacks on smart grid applications, in: Proceedings of the International Conference on Artificial Intelligence and Data Processing (IDAP), pp. 1−5, Sept. 2018.

[9] F. Aloul, A.R. Al-Ali, R. Al-Dalky, M. Al-Mardini, W. ElHajj, Smart grid security: threats, vulnerabilities and solutions, International Journal of Smart Grid and Clean Energy (2012) 1−6.

[10] P. Zhao, X. Chen, P. Yu, W. Li, X. Qiu, S. Guo, Risk assessment and optimization for key services in smart grid communication network, in: Proceedings of the IEEE International Symposium on Integrated Network Management, 2017.

[11] J.E.Y. Rossebo, R. Wolthuis, F. Fransen, G. Björkman, N. Medeiros, An enhanced risk-assessment methodology for smart grids, IEEE Computer 50 (4) (2017) 62−71. Apr.

[12] An. J. Rossebø, R. Wolthuis, F. Fransen, G. Björkman, Nuno Medeiros, An enhanced risk-assessment methodology for smart grids, IEEE Computer 1 (2017). Available from: https://doi.org/10.1109/MC.2017.106.

[13] M.Z. Gunduz, R. Das, Cyber-security on smart grid: threats and potential solutions, Computer Networks 169 (2020) 107094.

[14] A. Sanjab, W. Saad, I. Guvenc, et al. Smart grid security: threats, challenges, and solutions, (2016). arXiv: 1606.06992

[15] Ghosal, M. Conti, Key management systems for smart grid advanced metering infrastructure: a survey, IEEE Communications Surveys Tutorials 21 (3) (2019) 2831−2848.

[16] V. Dehalwar, R.K. Baghel, M. Kolhe, Multi-agent based public key infrastructure for smart grid, in: Proceedings of the Seventh International Conference on Computer Science & Education (ICCSE), 2012, pp. 415−418, Available from: https://doi.org/10.1109/ICCSE.2012.6295104.

[17] G. Drivas, A. Chatzopoulou, L. Maglaras, et al. A nis directive compliant cybersecurity maturity assessment framework (2020). arXiv: 2004.10411.

[18] M.A. Ferrag, L. Maglaras, S. Moschoyiannis, et al. Deep learning for cyber security intrusion detection: approaches, datasets, and comparative study, Journal of Information Security and Applications 50 (2020) 102419.

[19] A. Cook, H. Janicke, R. Smith, et al., The industrial control system cyber defence triage process, Computers & Security 70 (2017) 467−481.

[20] L.A. Maglaras, J. Jiang, Ocsvm model combined with k-means recursive clustering for intrusion detection in SCADA systems, in: Proceedings of the Tenth International Conference on Heterogeneous Networking for Quality, Reliability, Security and Robustness, 2014, 133−134.

[21] P.I. Radoglou-Grammatikis, P.G. Sarigiannidis, Securing the smart grid: a comprehensive compilation of intrusion detection and prevention systems, IEEE Access 7 (2019) 46595−46620.

CHAPTER 11

Blockchain-based secured payment in IoE

S.B. Gopal[1], C. Poongodi[2] and D. Nanthiya[3]
[1]Department of Electronics and Communication Engineering, Kongu Engineering College, Perundurai, Tamil Nadu, India
[2]Department of Computer Science and Engineering, Vivekanandha College of Engineering for Women, Thiruchengode, Tamil Nadu, India
[3]Department of Computer Technology-UG, Kongu Engineering College, Perundurai, Tamil Nadu, India

11.1 Introduction

Internet of things (IoT) includes the IoE as a subset. The IoT is built on the foundation of network intelligence, which is referred to as IoE. The Internet of Everything's (IoE's) spurpose is to transform data into actions, allowing for data-driven decision-making and introducing new capabilities and experiences. Machines to machines, machines to people, and technology–assisted people to people are all connected via the IoT. IoE is more complicated than IoT since it encompasses the IoD (Internet of Digital), IoH (Internet of Human), and IoT.

The Internet of things (IoT) is a collection of networked physical devices and things that collect and exchange data across wireless networks. The IoT is made up of two parts: the "Internet," which acts as the connection's backbone, and the "Things," which refers to real things and equipment. It connects the internet's power, data processing and analytics, and decision-making to physical items in the real world [1]. The IoT aims to build a network of linked objects and physical devices. Alternatively, you may create an ecosystem in which everything is interconnected. The IoT allows machines to communicate with one another. IoT is less advanced than IoE since it is considered a component of the wider IoE ecosystem. Wearable health monitoring, linked appliances, self-driving farming equipment, energy management systems that are more efficient, and smart surveillance are just a few examples. Bitcoin is one of the most well-known instances of Blockchain in operation [2]. This is a cryptocurrency (a type of digital cash). Bitcoin Atom (BCA) is a fork of Bitcoin

Smart Energy and Electric Power Systems
DOI: https://doi.org/10.1016/B978-0-323-91664-6.00009-7

that offers a genuinely decentralized alternative to trade cryptocurrencies with no trading fees or exchange hacks.

11.2 Blockchain

11.2.1 Working of blockchain

In cryptography, a private key and a public key are used. These keys help in the efficient execution of two-party transactions. These two keys are personal and are used to build a secure digital identification reference for each user. Secure identification is the most important aspect of Blockchain technology. In the bitcoin world, this identity is referred to as a "digital signature," and it is used to authorize and monitor transactions. The peer-to-peer network is paired with the digital signature; the digital signature is used by a large number of individuals acting as authorities to reach an agreement on transactions and other issues.

11.2.2 Decentralization

No one member (user) controls the transaction in a peer-to-peer network, and power is distributed across all network users, suggesting that no single stakeholder has the ability to hack, manipulate, halt, or shut down the chain of blocks [3]. This blockchain network is free of hackers and fraud owing to its distributed (decentralized) nature.

11.2.3 Integrity

All members in a blockchain-powered network have the right to make judgments; faith in the system is not imposed, but is entirely led by human intuition. The way each user is rewarded for their efforts, as well as the way this entire P2P network runs, demonstrates the high level of honesty.

11.2.4 Cryptography: fair to all

The primary premise of blockchain is to offer the user with a high level of security and authenticity [4]. It makes use of cryptography to maintain strict security and data integrity. The blockchain transaction process is advantageous and profitable for a trustworthy user, but it is extremely punitive for irresponsible users. The blockchain system is fair to all users who act well and rewards them accordingly. However, if you use it with the wrong intention, you will be punished.

11.2.5 Security

Because blockchain is a distributed network, there is no single point of failure, and no single person can destroy the entire chain by acting recklessly. Any damage caused by a single password breach will be limited to that single individual [5]. Unless you are a fool and leak your private key, your transaction over the network is extremely safe thanks to the blockchain PKI (public key infrastructure) encryption mechanism, for which no technology has a solution. As a consequence, you can trust Blockchain technology to conduct your transactions in a secure manner.

11.2.6 Inclusive

In its methodology, the blockchain is inclusive, allowing anybody to participate in the global economy without prejudice. Because a bank account is no longer required, Bitcoin allows anybody, rich or poor, to invest in their skills and participate in the global economy. They have the freedom to deal with anyone they choose without the need for a third party, and they can do it for free or for a very low transaction cost. Blockchain, in my opinion, is a link that will liberate and unite all people without discrimination. The major contribution of the blockchain to the world will be to enable everyone to participate in a global economy that is fair, safe, secure, and just to everyone.

11.2.7 Blockchain respects your privacy

Data privacy has grown crucial in a digital age when consumers and businesses trade online for a number of purposes, including purchasing, paying bills, and validating information [6]. Its strong hash key encryption is very secure, letting you to send and receive data over the internet while being anonymous. When you use the blockchain P2P network to conduct business, your identity is never disclosed [7]. As you become a member of the blockchain ecosystem, your rights and freedom become apparent and enforceable. The Smart Contract on the Blockchain is an excellent approach to carry out any agreement between two parties, as it allows transactions to take place only once certain predetermined benchmarks have been satisfied and both parties have consented.

11.2.8 Challenges in block chain

11.2.8.1 Scalability

The blockchain industry's capacity to handle a large number of users at once remains a hurdle. To conduct a single transaction, blockchain technology uses numerous sophisticated algorithms. The overall number of coinbase users was estimated to be 11.7 million as of October 2017. The typical transaction has grown considerably as more individuals have become accustomed to it. It had a significant impact on transaction processing speed since a larger number of individuals means more computers writing and accessing the network, resulting in a more cumbersome system overall.

11.2.8.2 Hackers and shadow dealing

The lack of governmental control in the blockchain business makes it a volatile environment and an easy target for market manipulation. For example, the famed one coin hoax, in which many investors lost money believing it to be the next breakthrough digital currency, turned out to be a Ponzi scheme. No matter how knowledgeable you are about cryptocurrencies, there is always the risk that your online wallet will be hacked or blacklisted by the authorities owing to unethical behavior.

11.2.8.3 Complex to understand and adopt

The complexity of blockchain technology makes it difficult for a layperson to comprehend and understand its benefits. Before going into this ground-breaking program, one must first read it and grasp the concepts of encryption and distributed ledger [8]. Another factor that makes blockchain difficult to implement is that, in comparison to the expenses associated with blockchain, financial institutions are capable of providing secure payment gateways and other services at reasonable pricing.

11.2.8.4 Privacy

Blockchain is an open ledger that can be viewed by anybody. In many circumstances, it is necessary, but when employed in a delicate context, it becomes a hazard [9]. Blockchain technology still has a long way to go before it is widely embraced. The ledger has to be redesigned in such a manner that it can only be accessed by those who are permitted to see it.

11.2.8.5 Costs

Blockchain is typically used to eliminate costs associated with third parties and intermediaries involved in the value transfer process. Despite its numerous benefits, blockchain technology is still in its early stages of development, making it challenging to integrate into current systems.

It changes into an expensive affair in general, preventing it from being adopted by both the government and commercial businesses.

11.2.8.6 Blockchain is still a distant dream

The blockchain technology, its benefits, and how it is reshaping the architecture of emergent technologies like InsurTech and others have market commentators enthralled. However, these obstacles are still difficult to overcome, and it will take some time before blockchain becomes an intrinsic component of all businesses.

11.2.9 Application of block chain

11.2.9.1 Money transfers

The initial notion behind blockchain technology's creation is still a fantastic use. Traditional methods of money transfer may be more expensive and time consuming than using the blockchain. This is particularly true in the case of cross-border transactions, which are notoriously delayed and costly [10]. Money transfers between accounts can take days in the existing US banking system, but it takes minute to complete the transaction.

11.2.9.2 Financial exchanges

Over the last several years, a slew of firms have sprung up to provide decentralized bitcoin exchanges. When it comes to exchanges, blockchain enables for speedier and less expensive transactions. Furthermore, because a decentralized exchange does not force investors to deposit their funds with a centralized authority, they have more control and security. While bitcoin is the primary focus of blockchain-based exchanges, the concept might be extended to other traditional assets as well.

11.2.9.3 Insurance

Customers and insurance providers can benefit from more transparency by using smart contracts on a blockchain. Customers would be prevented from filing repeated claims for the same occurrence if all claims were recorded on a blockchain. Furthermore, the use of smart contracts can expedite the payment procedure for claimants.

11.2.9.4 Real estate

To verify financial facts and ownership, as well as transfer deeds and titles to new owners, real estate transactions need a large amount of documentation. Real estate deals are being recorded using blockchain technology can improve the security and accessibility of ownership verification and transfer. This can help you save time and money by speeding up transactions and reducing paperwork.

11.2.9.5 Data storage

When blockchain technology is used in conjunction with a data storage system, it may improve security and integrity. Because data is maintained decentralized, Hacking into and wiping out all of the data on the network will be more difficult, whereas a centralized data storage provider may only have a few points of redundancy. It also suggests that data is more available because access is not restricted to the actions of a single organization. In certain circumstances, storing data on blockchain might be as less expensive as other types of storing data.

11.2.9.6 Gambling

Blockchain might be used by the gaming industry to benefit gamers in a variety of ways. The transparency that operating a casino on the blockchain delivers to potential players is one of the main advantages. Bettors can see that the games are fair and that the casino pays out because every transaction is recorded on the blockchain. Furthermore, using blockchain eliminates the need for personal information such as a bank account, which may deter some would-be gamblers. Because users may wager anonymously and the decentralized network is not subject to government closure, it provides a workaround for regulatory constraints.

11.3 Internet of everything (IoE)

IoE proposes a scenario in which billions of items are connected via public or private networks utilizing standard and proprietary protocols to detect, measure, and analyze their state as shown in Fig. 11.1.

Need for IoE

11.3.1 Benefits of the Internet of things

- More information improves overall conclusions.
- The ability to keep track of and monitor events
- Automate your tasks to reduce your workload.
- Saves money and resources, resulting in increased efficiency.
- Better quality of life

Figure 11.1 Internet of everything.

11.3.2 Challenges in privacy and security

IoE is a system with several domains. IoE is made up of a large number of connected devices and services that share data. Every domain has its own set of norms for security, privacy, and trust [11].

The following are some of the current problems in IoE securities: privacy of user and protection of individual data in

- Authentication and administration of identities
- Integrating policies and managing trust
- Access control and authorization
- Point-to-point security
- Security solution that is resistant to attacks

11.3.3 Communication between Machine to Machine

Machine to Machine (M2M) is a communication technology that allows wireless and wired equipment to connect with one another. In industrial automation, M2M connections are commonly utilized for machine monitoring and instrumentation. The "Internet of verything" (IoE) is a collection of people, processes, data, and objects that collaborate to improve the utility and relevance of networked interactions. IoE encompasses both M2M and IoT technologies, and its pervasiveness allows it to be utilized to achieve a wide range of goals for a wide range of people, including first responders as shown in Fig. 11.2.

IoE has taken the role of the IoT. Nowadays, security in an IoE network is critical. The key hurdles in IoE are security and privacy concerns. The numerous IoE security standards are beneficial in automating systems in various areas such as home automation, smart cities, smart agriculture, and so on.

11.3.4 Application

11.3.4.1 Smart city

To collect and analyze data, smart cities rely on IoE devices such as connected sensors, lights, and meters. Cities then utilize the information to enhance their infrastructure, public utilities, and other areas of their operations [12]. The Internet of Things (IoE) solutions available in the smart city sector serve to handle a variety of city-related issues, such as traffic congestion, air and noise pollution reduction, and city safety.

11.3.4.2 Smart grids

The smart grid is part of an IoT architecture that may be used to monitor and control everything from lighting to traffic signals, traffic congestion, parking spots, road alerts, and early detection of things like earthquake-induced power surges [13]. The smart grid distributes transmission lines, smart meters, distribution automation, substations, transformers, sensors, software, and more to companies and families around the city [14].

11.3.4.3 Health

In the healthcare industry, the most recent IoE trends are being used. IoE software supports patients in decreasing health-related dangers and hospital

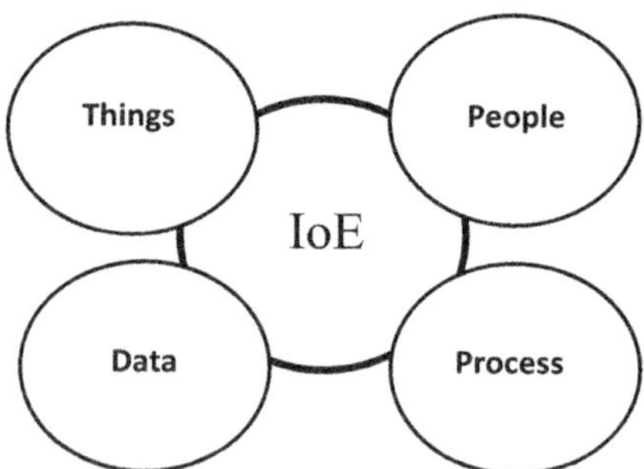

Figure 11.2 IoT with M2M. *IoT*, Internet of things; *M2M*, Machine to Machine.

expenses by gathering patient information and analyzing data using cloud services to exchange data sources.

11.3.4.4 Retail

Retail enterprises may now evolve into smart shops that collect data on consumers' preferences, requirements, and behaviors in real time thanks to the IoT. This allows merchants to understand client behavior and deliver the items and services they desire.

11.3.4.5 Smart supply chain

For a number of years, supply networks have been growing smarter [15]. Tracking things while they are on the road or in transit is one of the most popular services, as is supporting suppliers with inventory information exchange. Factory equipment with embedded sensors may send data about many factors, such as pressure, temperature, and machine usage, using an IoE enabled system [16]. The IoE system may also enhance performance by processing operations and changing equipment settings.

11.3.4.6 Smart farming

The IoE can monitor a vast number of farming activities and also lot of livestock that farmers goes through it, all of them which can transform the way farmers operate day to day.

11.4 Secure payments in IoE

11.4.1 IoE with blockchain: security challenges

Despite the fact that blockchain technology greatly improves the dependability and concerns are always developing [17]. This article examines the security flaws that have arisen as a result of the confluence of blockchain and IoT, as well as possible remedies [18].

The following security problems in implementing dispersed IoT are highlighted in this article: (1) Network security as well as communication; (2) identity management and verification; (3) dispersed consensus protocol with good dependence; (4) distributed cooperation and establishment of trust; (5) transferring data privacy and security (P2P) mode which is peer to peer method, which allows nodes to interact with each other autonomously without the use of a central server platform, is used heavily in the architecture of IoT. One of the important technologies for resolving the aforesaid security issue is cryptographic algorithms. In the IoT, (PKIs) make identity association and verification easier as shown in Fig. 11.3. At the moment, the most widely used PKIs are (CAs) and web of trust based on somewhat excellent privacy.

11.4.2 A secures energy transaction prototype

The communication related infrastructure, 5G components are discussed in this project as shown in Fig. 11.4. Here 5G-V2X network model for the EVBlocks trading scheme is shown. We analyze mm Wave channel-based communication in the 5G-V2X ecosystem, which works in the thirty to three hundred GHz range. Nodes N1 and N2 connect with

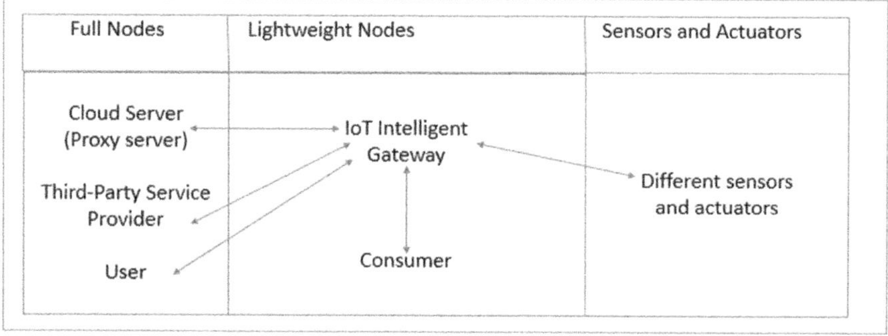

Figure 11.3 IoE with blockchain.

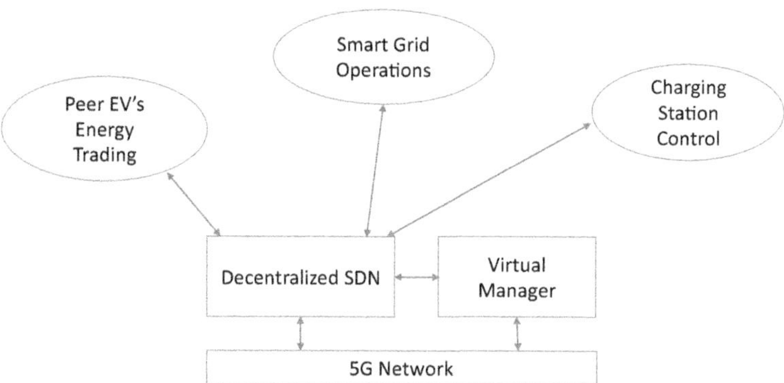

Figure 11.4 Secured energy transaction.

each other using transmitting and receiving antennas in C–V2X. The SDN-leveraged CBC resource-trading model and the accompanying problem formulation are seperated. Layer from 1, 2 and 3 are the layers of the resource-trading model [19]. EVs communicate with CS situated GS with the necessary frequency range for energy exchange. Micro-grids of small in size, known as location aggregators (LAGs), are located within each GS to help with load-balancing.

11.4.3 Internet of entities (IoE): a blockchain-based distributed paradigm to security

It is an effective workflow for the localization of people and things that takes advantage of the vast number of existing wireless-based devices to overcome the limitations while maintaining user privacy.

A tracker device is used for detection, a distributed ledger is used for recording the collected data [20] as shown in Fig. 11.5. With a focus on the characteristics of current wireless technologies capable of performing these activities. It formalises the structures of the data and procedures involved in software is capable of combining information connected to the participating entity and devices used for tracking, resulting in a structure of data that represents the statistics to be stored on the distributed ledger. Localization of elements: it thoroughly outlines the work carried out, introducing certain methodologies and a set of rules for localization targeted at directly or indirectly exploiting the information accessible on the blockchain (Fig. 11.6).

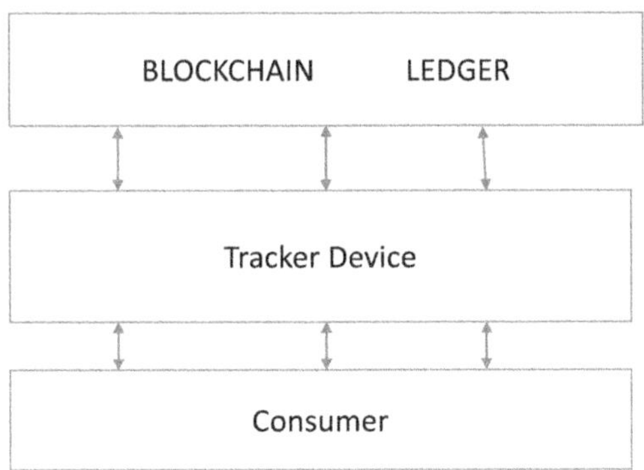

Figure 11.5 Blockchain-based distributed paradigm.

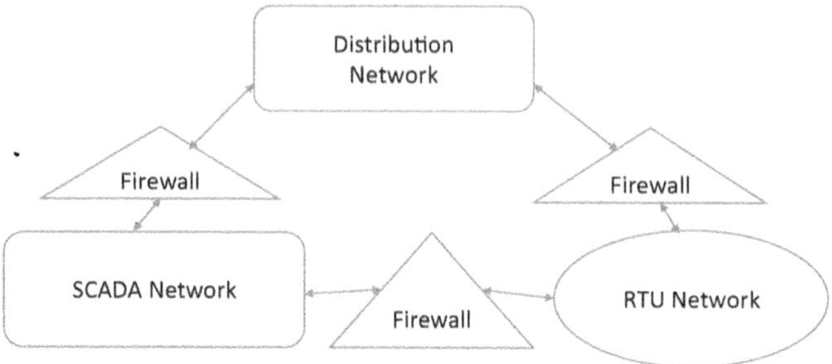

Figure 11.6 Secured blockchain prototype.

11.4.4 A secured blockchain prototype

The smart RTUs of SCADA are used to gather, transmit, and receive customized signal data between the communication and physical layers. The RTU-nodes are used to connect the communication paths. Only authorized individuals with communication access can confirm data collecting and processing for reports or signaling of various processes used in the production environment as shown in Fig. 11.6. The proposed decentralized data security solution relies heavily on advanced metering infrastructure (AMI).

11.4.5 Role of software in securing Internet of things/IoE

The purpose of this article is to analyze the current importance of security needs in modern cultures.

The above mentioned methods were used to collect primary and secondary data: (1) obtaining a large amount of unidentified practical data about software security vulnerabilities from a cyber intelligence company that works in a many countries; (2) conducting interviews that are informal with two well-known cyber intelligence workers [21] as shown in Fig. 11.7.

11.4.6 Role of IoE/Internet of things in vehicular networks

This research will aid researchers in gaining a thorough understanding of the taxonomy of automotive network security challenges, which can then be used to develop novel solutions. This knowledge will also aid in the development of new research avenues, which will aid in the design of new tactics for dealing with attacks more effectively.

Cars and electric vehicle charging grids are all part of VANETs [3,5,22]. In VANETs, there are two types of communication: vehicle-to-vehicle (V2V) and vehicle-to-infrastructure (V2I), as well as infrastructure-to-vehicle

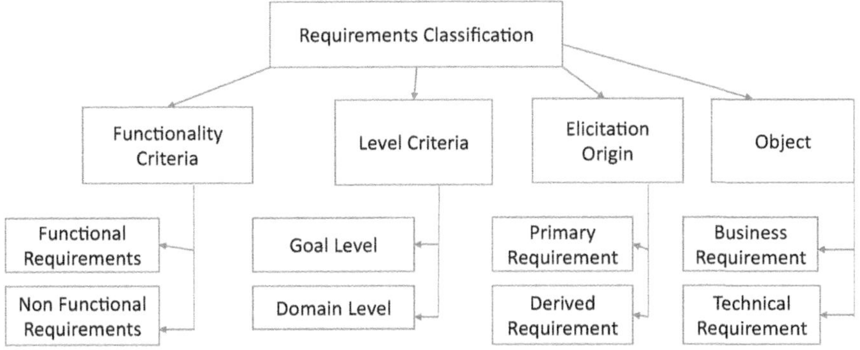

Figure 11.7 Secured blockchain prototype.

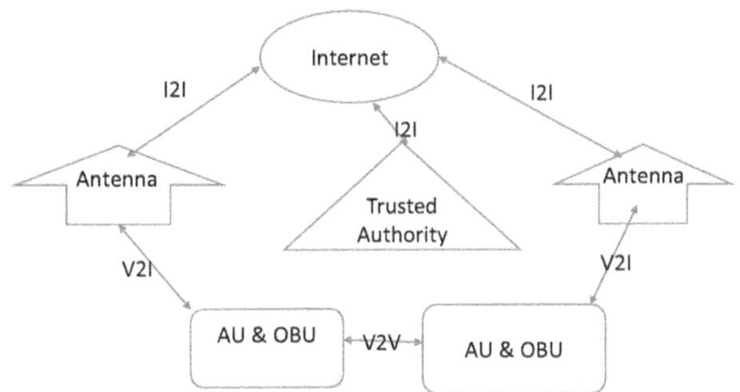

Figure 11.8 IoT in vehicular networks. *IoT,* Internet of things.

(IV) (I2V). Communication between roadside units and base stations is known as infrastructure-to-infrastructure (I2I) as shown in Fig. 11.8. As a backbone, they rely on the internet.

11.4.7 IoE with privacy

Rootstock (RSK), Sidechain Overlay Block Manager (OBM), and 2 Way Peg are the names of our proposed IoE architecture (WP). OBM is a company that manages an integrated blockchain that operates on one or more sidechains and is in charge of generating, verifying, and storing individual transactions. A two-way peg (2WP) protocol is one that allows a bitcoin to be transferred from one blockchain to another inside a third-party trust as shown in Fig. 11.9.

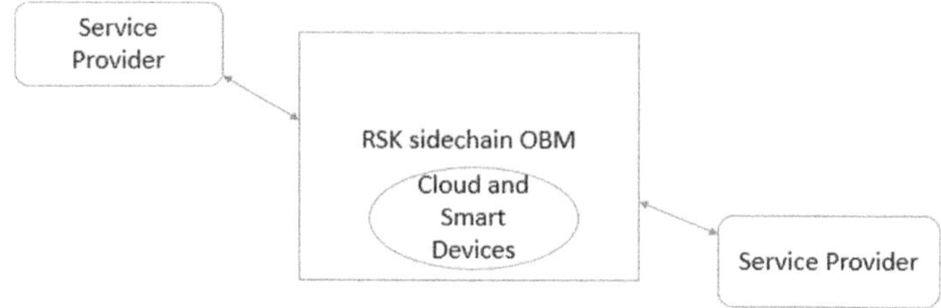

Figure 11.9 IoE with privacy.

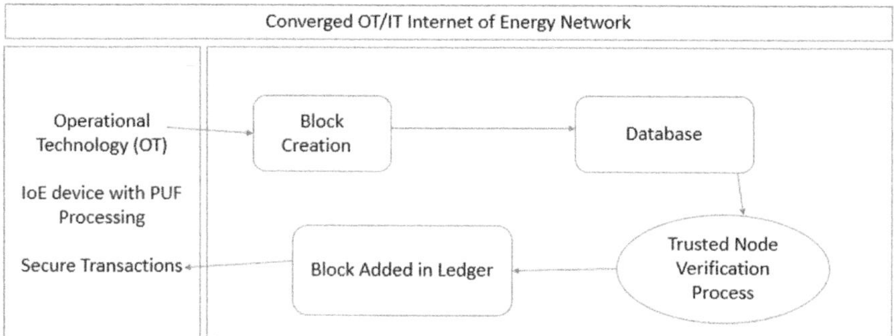

Figure 11.10 PUF enabled secured blockchain. *PUF*, Physical unclonable function.

OBM is a company that manages an integrated blockchain that runs on one or more sidechains and is in charge of generating, verifying, and storing individual transactions. A 2WP protocol is one that allows a bitcoin to be transferred from one blockchain to another inside a third-party trust. We wish to analyze existing 2WP approaches like Federation, Sidechain, and Drivechain, as well as hybrid ways, to determine which is best suited for the RSK platform's 2WP at each level of adoption. Rootstock was developed and tested to give two methods for constructing and confirming SPV proofs. The Rootstock 2WP smart contract creates two separate indicators to lock and unlock cryptocurrencies like bitcoin.

11.4.8 Physical unclonable function enabled secured blockchain

This paper provides a thorough examination of the technological features of blockchain and physical unclonable functions (PUFs). It defines a developing blockchain idea that uses PUFs to incorporate security in hardware primitives to address bandwidth, integration, scalability, latency, and energy needs for Internet–of–Energy (IoE) devices as shown in Fig. 11.10.

Blockchain, which is based on Proof-of-PUF verification, data stored which is encrypted with same PUF-derived keys and attributes, providing immutable assurance that data have not been changed with, as well as traceability and simple auditing capabilities. The shuffling of the PUF technology, as mentioned in the preceding sections, is crucial to the PUF Chain's defence. The amount of 0 and 1 s should be the same for PUF answers to be highly consistent and impervious to the cyber assaults.

References

[1] M. Moness, A.M. Moustafa, A survey of cyber-physical advances and challenges of wind energy conversion systems: prospects for internet of energy, IEEE Internet of Things Journal 3 (2016) 134−145.

[2] A. Verma, et al., The security perspectives of vehicular networks: a taxonomical analysis of attacks and solutions, Applied Sciences 11 (10) (2021) 4682.

[3] N.Z. Aitzhan, D. Svetinovic, Security and privacy in decentralized energy trading through multisignatures, blockchain and anonymous messaging streams, IEEE Transactions on Dependable and Secure Computing 15 (5) (2018) 840−885.

[4] M.T. Hammi, et al., Bubbles of trust: a decentralized blockchain-based authentication system for IoT, Computers & Security 78 (2018) 126−142.

[5] R. Roman, J. Zhou, J. Lopez, On the features and challenges of security and privacy in distributed internet of things, Computer Networks 57 (10) (2013) 2266−2279.

[6] Z. Liu, et al., On emerging family of elliptic curves to secure internet of things: ECC comes of age, IEEE Transactions on Dependable and Secure Computing 14 (3) (2017) 237−248.

[7] A. Satsiou, L. Tassiulas, Reputation-based resource allocation in P2P systems of rational users, IEEE Transactions on Parallel and Distributed Systems 21 (4) (2010) 466−479.

[8] A. Kiayias et al., Ouroboros: A provably secure proof-of-stake blockchain protocol, in Proc. Annu. Int. Cryptol. Conf., 2017, pp. 357−388.

[9] Z. Lu, et al., A privacy-preserving trust model based on blockchain for VANETs, IEEE Access. 6 (2018) 45655−45664.

[10] Y. Zhang, R.H. Deng, X. Liu, D. Zheng, Blockchain based efficient and robust fair payment for outsourcing services in cloud computing, Information Sciences 462 (2018) 262−277.

[11] S.B. Gopal, C. Poongudi, D. Nanthiya, A. Jeevanantham, D. Sumithra, P. Tharani, et al., Secured smart home using blockchain technology, Gedrag & Organisatie Review (2020)ISSN:0921-5077.

[12] X. Fang, S. Misra, G. Xue, D. Yang, Smart grid—the new and improved power grid: a survey, IEEE Communications Surveys & Tutorials 14 (4) (2012) 944−980.

[13] N. Bui, A.P. Castellani, P. Casari, M. Zorzi, The internet of energy: a web-enabled smart grid system, IEEE Network 26 (2012) 39−45.

[14] V.C. Gungor, B. Lu, G.P. Hancke, Opportunities and challenges of wireless sensor networks in smart grid, IEEE Transactions on Industrial Electronics 57 (10) (2010) 3557−3564.

[15] Y. Yan, Y. Qian, H. Sharif, D. Tipper, A survey on smart grid communication infrastructures: Motivations, requirements and challenges, IEEE Communications Surveys & Tutorials 15 (1) (2014) 5−20 (2013).

[16] F. Mwasilu, J.J. Justo, E.-K. Kim, T.D. Do, J.-W. Jung, Electric vehicles and smart grid interaction: a review on vehicle to grid and renewable energy sources integration, Renewable & Sustainable Energy Reviews 34 (2014) (2014) 501−516.

[17] L. Wei, J. Wu, C. Long, Y.B. Lin, The convergence of ioe and blockchain: security challenges, IT Professional 21 (5) (2019) 26−32.

[18] S.B. Gopal, D. Nanthiya, P. Keerthika, S.B. Kayalvizhi, T. Raja, R. Snega Priya, SVM Based DDoS Attack Detection in IoT Using Iot-23 Botnet Dataset, Innovations in Power and Advanced Computing Technologies (I-PACT) (2021) 1−7.

[19] P. Bhattacharya, S. Tanwar, U. Bodkhe, A. Kumar, N. Kumar, EVBlocks: a blockchain-based secure energy trading scheme for electric vehicles underlying 5G-V2X ecosystems, Wireless Personal Communications (2021) 1−41.

[20] R. Saia, S. Carta, D.R. Recupero, G. Fenu, Internet of entities (IoE): a blockchain-based distributed paradigm for data exchange between wireless-based devices, SENSORNETS (2019) 77−84.

[21] F.F.S. Flores, S.R. de Lemos Meira, Ethical software engineering: a critical review about software engineering in face of security requirements in the IoT/IoE society, in: 2021 IEEE International Systems Conference (SysCon), 2021.

[22] M. Suarez-Albela, et al., A practical evaluation of a high-security energy-efficient gateway for IoT fog computing applications, Sensors 17 (9) (2017). 2017, Art. 1978.

Index

I

K

L

M

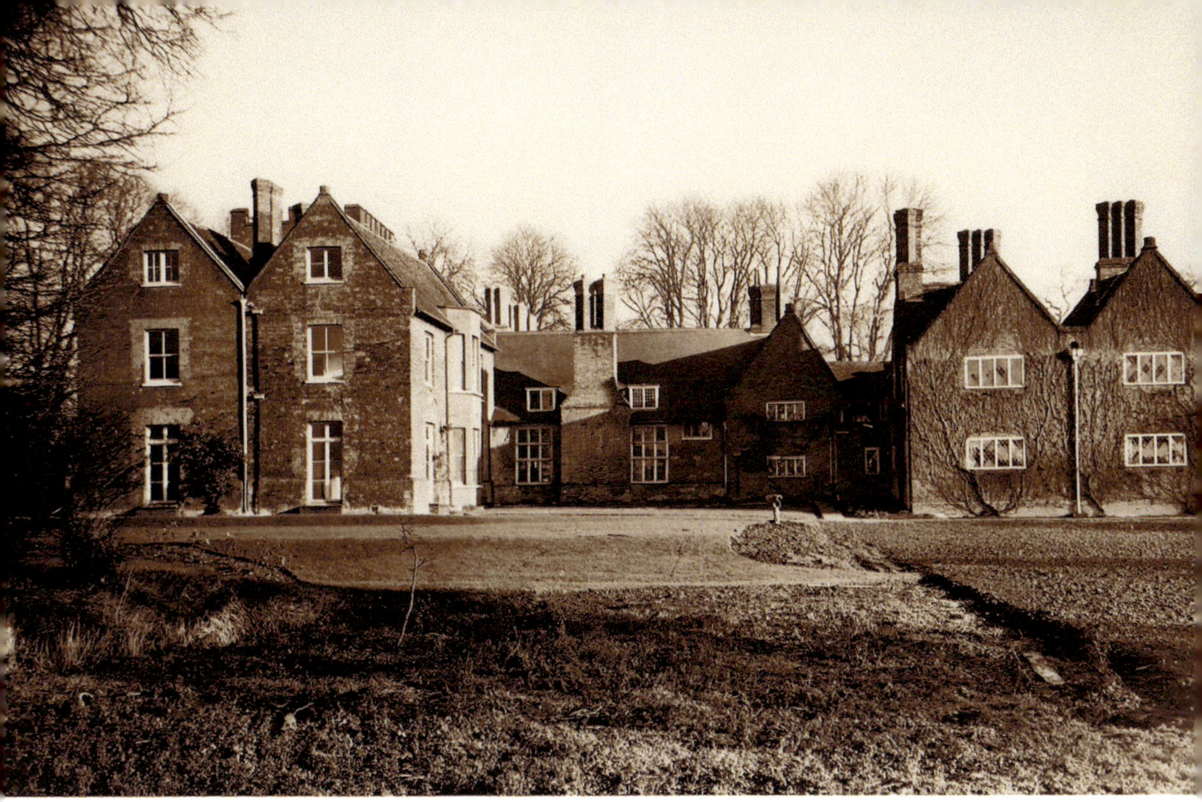

Letchworth Hall Hotel

The building today has had several significant extensions and renovations, but the bulk of the original buildings retain their character and charm, and it is still a popular and thriving hotel, with restaurant and conference facilities.

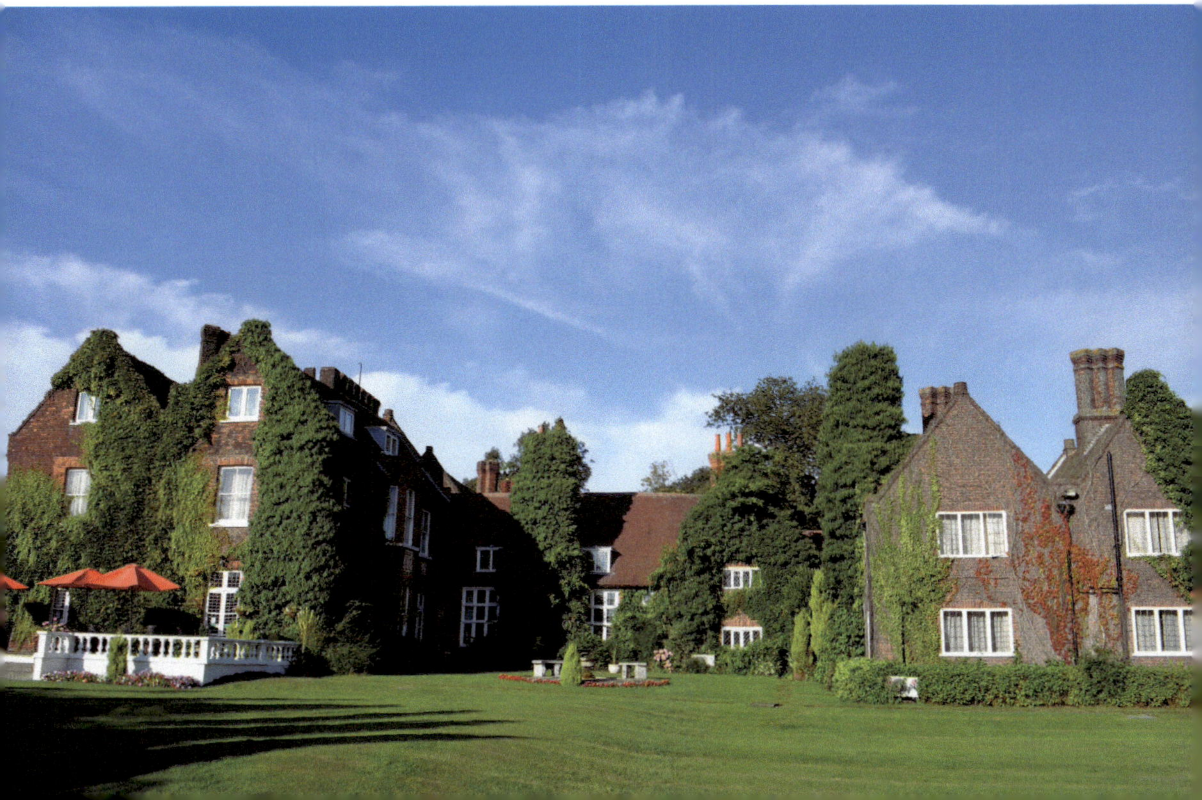

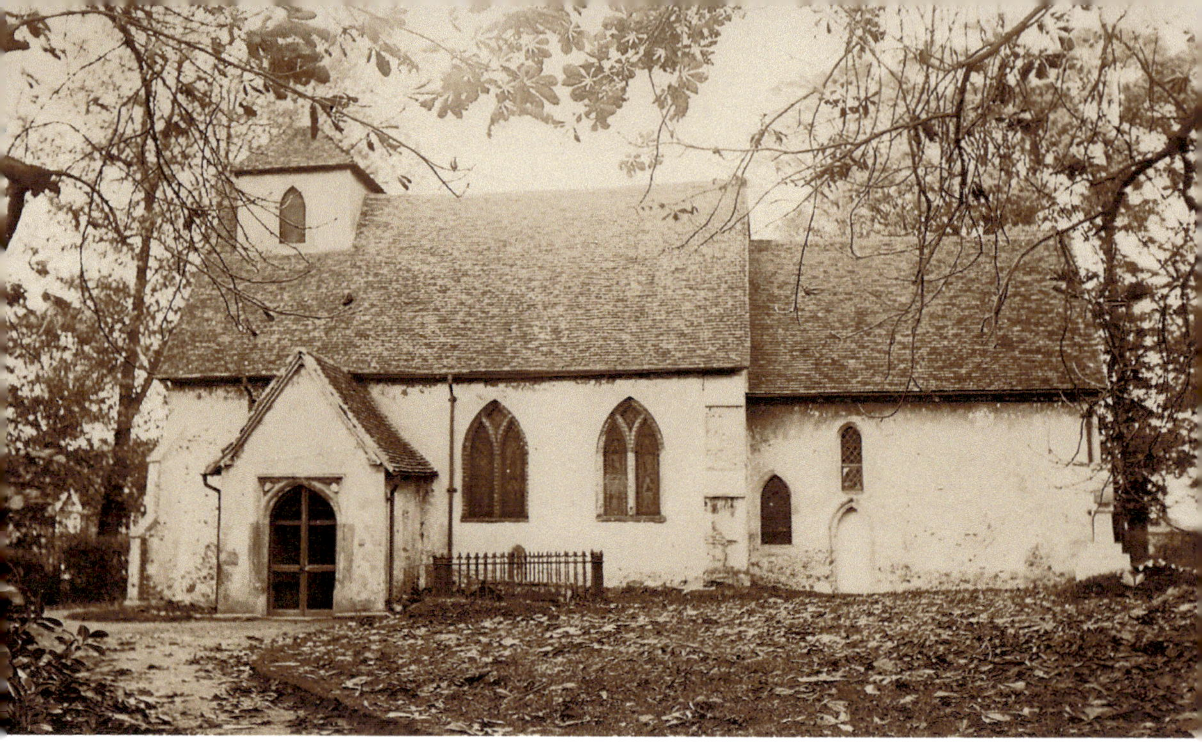

St Mary's Church

The Church of St Mary the Virgin, in the original hamlet of Old Letchworth dates from at least the late twelfth century, although it existed in some form as early as the Domesday book in 1086. Indeed, building material found in the churchyard suggests that there may have been an earlier Saxon structure on the site. The porch was added during the fifteenth century and the belfry in around 1500 (housing a much older bell).

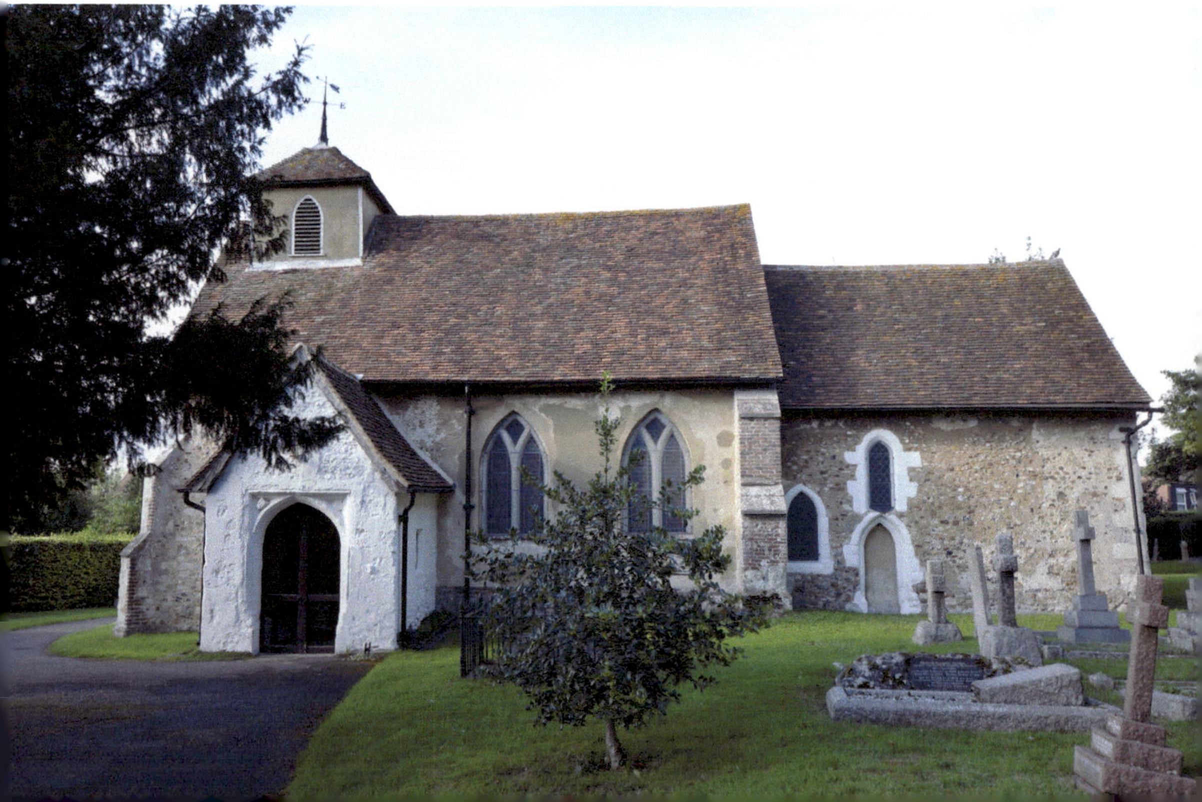

St Mary's Church

The diminutive size of the church – its interior is just 60 feet long – reflects the remoteness and small size of Letchworth Manor. It remained the parish church for Old Letchworth until 1967 when St Michael's was built on Broadway.

S 16726   LETCHWORTH CHURCH.

The Heart of a Crusader?

Among its notable internal features, the church contains a small recumbent effigy of a crusader, the knight Sir Richard de Mountfichet. His heart is reputed to have been buried beneath the church. The building has changed very little over the centuries and retains its quiet charm today.

Willian

Located close to Old Letchworth, in the south of the estate is the village of Willian. Dating back to Saxon times, it has retained a strong village character, despite the Garden City springing up around it. It has a population of about 300 or so, and much of the village was rebuilt in the 1860s by the then landowner Charles Hancock, including the group of cottages pictured here across the village's picturesque pond.

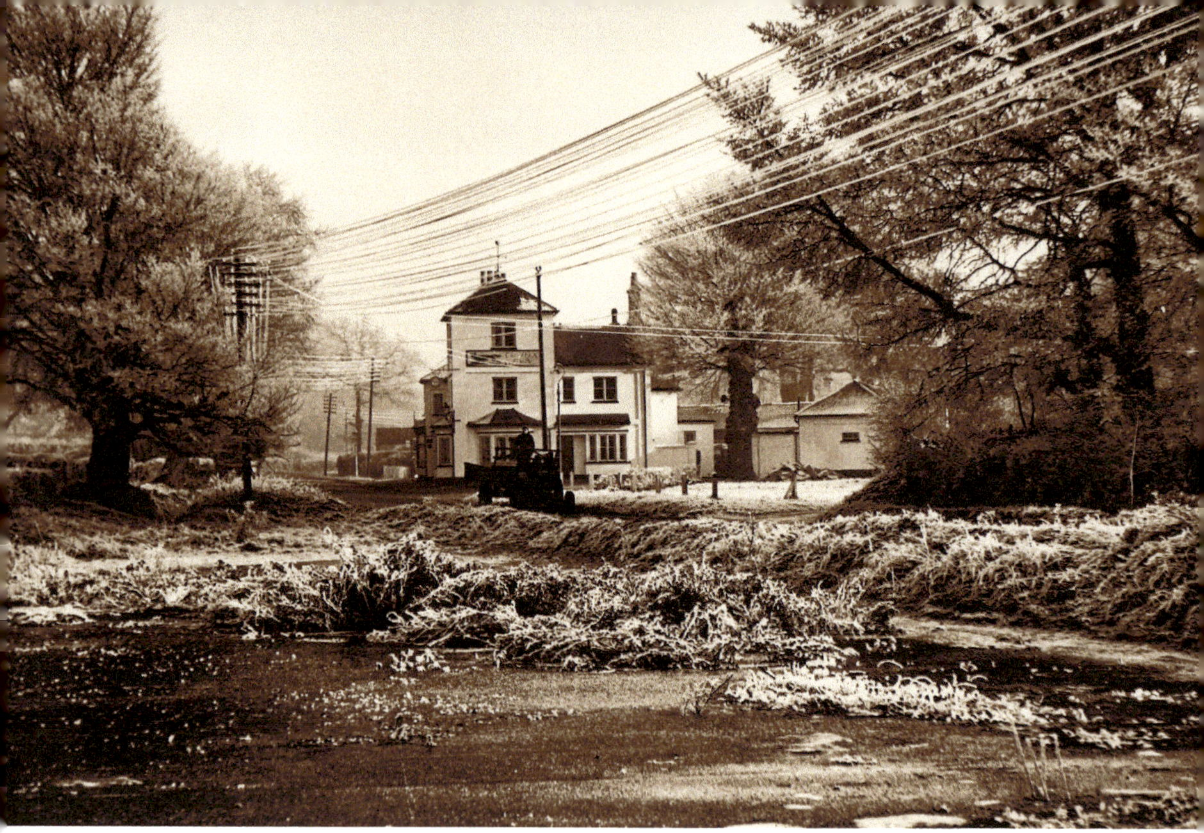

Willian Pond

This more classic view the other way back across Willian Pond takes in the village's main street, with a post office and The Fox public house (opened in 1902), with All Saints Church towering behind them.

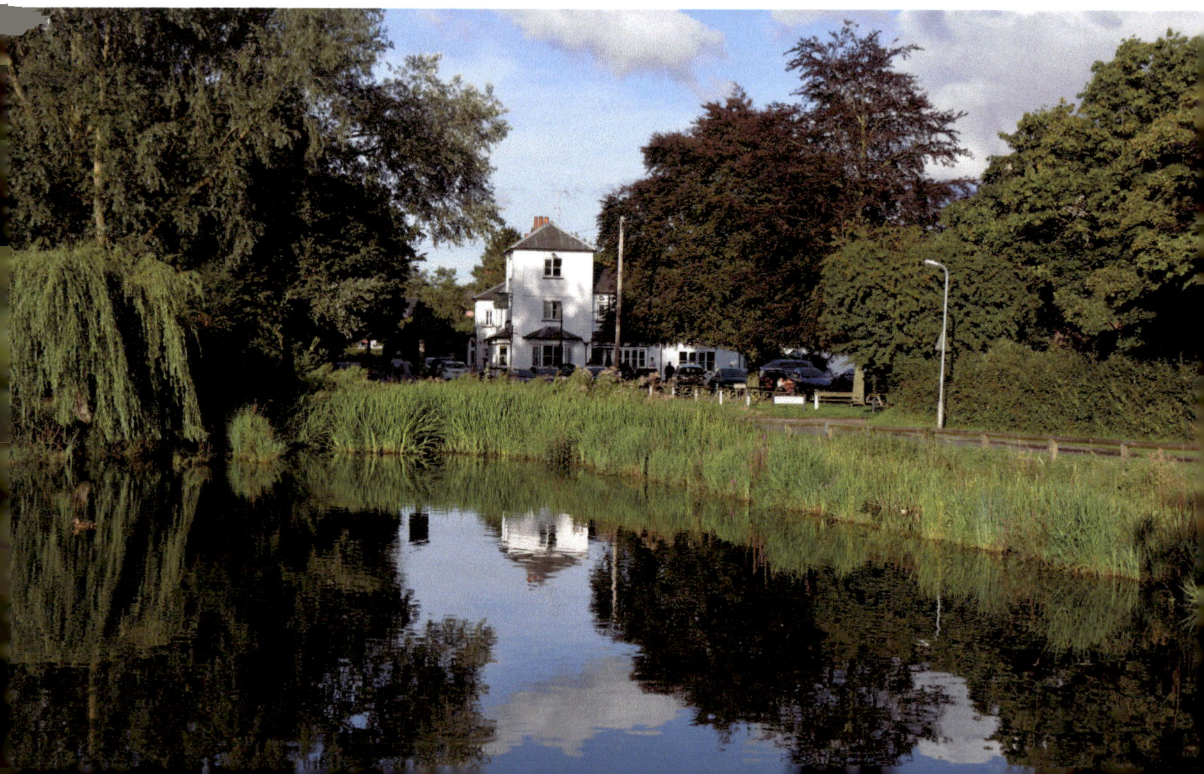

Norton

To the north of the Letchworth Garden City estate lies the historic village of Norton, parts of which date back to the eleventh century, with evidence of settlement here in Roman, Iron Age and even the prehistoric eras. This classic often-photographed view, along the village's main road, takes in early cottages, The Three Horseshoes public house, St Nicholas Church and its attached school.

The Three Horseshoes

The Three Horseshoes is the social hub of the village, much extended in the years since the image above. The photographer, prolific local man Julian Tayler, produced many evocative postcards of Letchworth in the 1920s and '30s, often with their title inventively incorporated into the image, as here with the word 'Norton' appearing under the car on the left.

Lych Gate, Norton St Nicholas' Church
The Church of St Nicholas was the parish church for the village for over 850 years, until St George's Church was built in Letchworth Garden City. The church itself dates back to 1119, and it was extended and the tower added in the mid-1400s.

Norton St Nicholas' Church

This Julian Taylor postcard of the church – title lurking in the shadows beneath the tree in the foreground – was taken in the 1920s. The trees that provided a picturesque accompaniment to the church now dominate the church and churchyard.

Alpha Cottages

The first new houses to be built in the Garden City were these on the main Baldock-Hitchin Road, which were known as 'Alpha Cottages' and completed in 1904. Above, the labourers have downed tools to pose for a photograph of the progress being made in their construction. The houses have changed little in over a century since their construction.

'Letchworth Look' Housing

Early housing followed building regulations set out by Parker and Unwin, who approved or rejected designs and while many houses were designed by themselves or their assistants. Thus a distinctive early 'Letchworth look' emerged, with roughcast walls, gables, pitched roofs with red tiles, and green detailing on the doors, water butts, and (often dormer) windows. This early cul-de-sac, Rushby Walk, clearly demonstrates many of those features.

Rushby Mead

Running for some two miles or so, Rushby Mead is populated entirely with this archetypal style of housing, predominantly designed for factory workers and their families, and mostly today owned (and beautifully maintained) by a local housing association, The Howard Cottage Society, which was founded in 1912.

Rushby Mead

More classic early 'Letchworth look' housing on Rushby Mead. These featured a tiled gable, and decorative tile features above each front doorway. The road names in Letchworth are often deliberately rustic – Meads, Dales, Hills and Views – an attempt to evoke a romantic rural idyll, far from the squalor and overcrowding of the roads, streets and avenues of some of the towns of Victorian England.

Westholm

Here on Westholm Green, houses are arranged in groups around a communal green space in one of Parker and Unwin's favourite layout plans, the 'village green' design, a nod to the traditional medieval greens in the villages of old England. The effect is to set back the houses from the main road, successfully providing a distinctly rural feel.

RED HAWTHORN NORTON ROAD.

No. 158 Wilbury Road

This unique building was one of the world's first pre-fabricated houses. It was part of the 1905 Cheap Cottages exhibition, a housing competition which aimed at proving that good quality houses could be built for £150 or less using new and innovative materials and methods. Made mainly of concrete, with fittings fixed directly onto the walls, it was designed by the Liverpool City Surveyor, J. A. Brodie, and slotting together on-site like a 3D jigsaw, in under thirty-six hours.

Noah's Ark Cottages

Letchworth leapt at the opportunity to host the 1905 exhibition, which ran from June to September and provided over 120 houses. It proved a real publicity boost for the fledgling town, with national press coverage and more than 60,000 people visiting the town. These cottages on Birds Hill, designed by V. Dunkerley, achieved the economy in construction through the use of tiles in the mansard roof on the upper floor, giving them their distinctive appearance, and hence their nickname.

The Quadrant

These houses on The Quadrant were also entries in the competition; the house on the right, No. 8, was designed by architect Lionel Crane, the son of renowned illustrator, Walter Crane. It was furnished in 'artistic but inexpensive style' by Heals & Son, of Tottenham Court Road, London, and Heals later produced a 'Letchworth' range of furniture based on their designs for this 'show home' in the exhibition.

Cross Street

The photograph above of further exhibition cottages on Cross Street taken in 1905 illustrates very clearly the barren and unfinished nature of some of Letchworth's early roads. Early residents wrote poems about the perilous mud on Letchworth's unfinished roads, and one of them was even called Muddy Lane (which still exists) in recognition of the fact.

Cross Street

The view back the other way, taken from Icknield Way, gives a clearer view of these three exhibition entries. The house in the middle, No. 4 Cross Street, was designed by the architect Gilbert Fraser for the Concrete Machinery Company Ltd. The walls were built from concrete bricks made by the companies 'portable "pioneer" hand-power machine', which made a minimum of 150 blocks a day.

Nook Cottage

So named after its inglenook fireplace, No. 2 Cross Street was one of the most popular exhibition cottages. All the rooms were on one floor to simplify construction and avoid costly scaffolding. It was designed to be durable, fire-proof and cost the same as a timber cottage to build. Far from being a timber cottage, it used modern materials, dispensing with bricks and mortar in favour of a metal frame filled with seven-inch-thick concrete for the walls.

Exhibition Road

These cottages on Exhibition Road, now renamed Nevells Road, give an idea of the range of different sizes and types of houses. In the foreground, No. 212 was made by the Cement Products Company, built of pre-cast hollow concrete using the Bowen Hollow Block machine. According to the company, these blocks gave 'a perfect circulation of air and render the building warm in winter and cool in summer'. Its design was described as 'in the Canadian style entirely'.

Nevells Road and The Skittles Inn

This view looking from Cross Street over towards Exhibition (now Nevells) Road, offers a glimpse, on the left, of one of Letchworth's most well-known buildings. The Skittles Inn, which today (since 1923) is an adult education centre and small theatre venue called 'The Settlement'.

'The Pub With No Beer'

Built in 1907 and designed by Parker and Unwin, it was originally called The Skittles Inn, the town's famous 'pub with no beer'. The citizens of Letchworth voted regularly on 'the licence question', i.e. whether there should be a pub in the town, and regularly voted narrowly against. However, the town needed a social hub, and so The Skittles Inn was built, which offered all of the social aspects of a public house but without the alcohol.

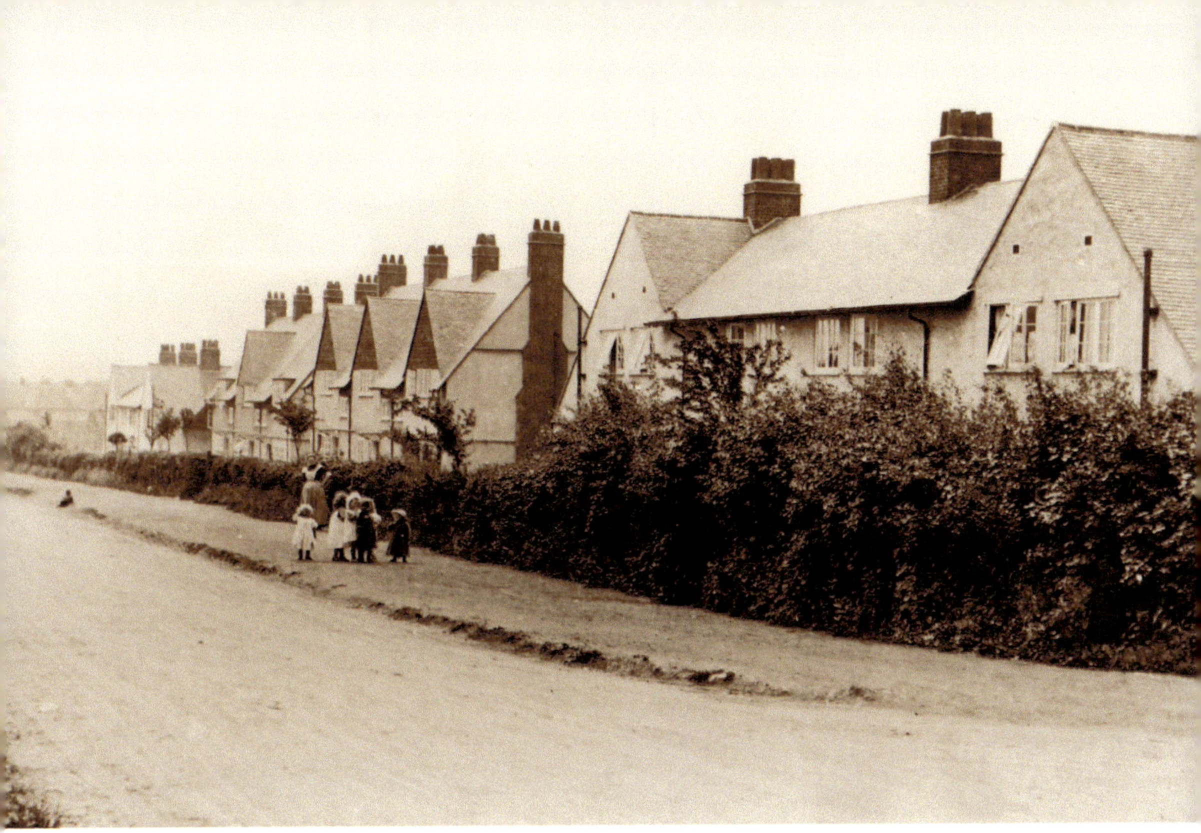

Green Lane

Houses on Green Lane – a road name aimed at evoking the romantic rural idyll if ever there was one, and still suitably verdant today.

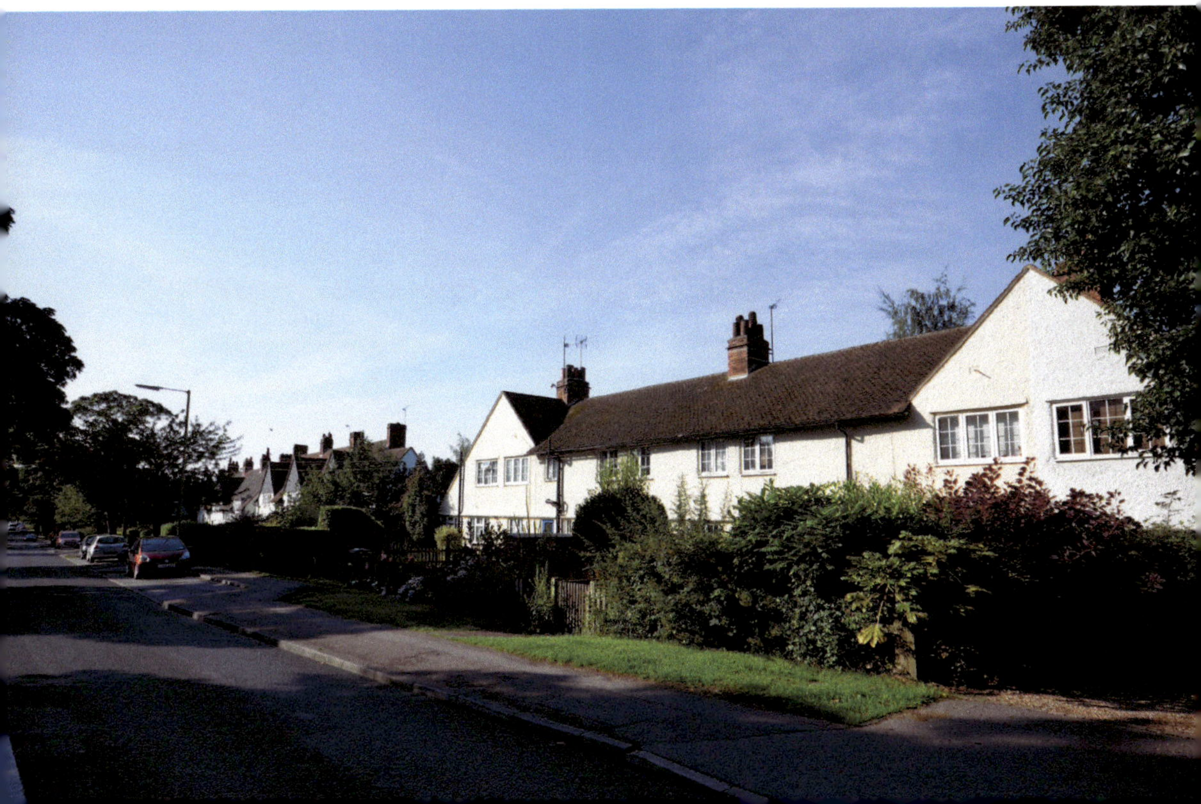

Birds Hill

This group of houses on the corner of Ridge Road and Birds Hill was built in 1906 as the first part of the Pixmore Hill estate which provided workers with housing within walking distance of, but screened off with trees from, the industrial area.

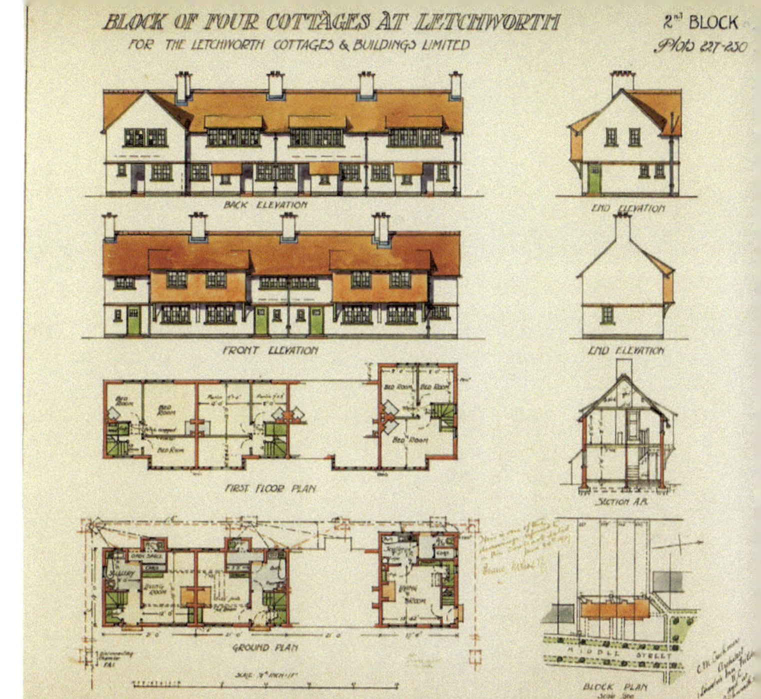

Lytton Avenue

These houses on Lytton Avenue were designed by Courtenay M. Crickmer as part of the 1907 Urban Cottages & Smallholdings Exhibition, a follow-up to the 1905 competition, which put the emphasis on groups of houses rather than individual cottages.

Meadow Way

Houses on Meadow Way, one of Letchworth's typical long, tree-lined residential streets. Each of the streets in Letchworth Garden City is lined with a different variety of trees.

Meadow Way Green

Designed by Crickmer and built by the Howard Cottage Society in 1914, this group of houses was a unique co-operative scheme for professional single business women. Midday meals were taken in a communal dining room, with tenants taking turns to purchase food and devise a menu, which was prepared by a cook. In addition, the landscaped front garden area was communally maintained.

Common View

An unusual unbroken terrace of workers' cottages, these houses on Common View were built by Letchworth Cottages & Buildings Ltd. Although economically built, each cottage still provided three bedrooms, although the bath was in the scullery and the toilet was outside. Located just to the north of the industrial area along the railway line, a network of alleys link Common View to the factories, providing an easy walk to work.

Garden City Tenement Factory

One of the earliest factories in the town, which played host to a whole number of different businesses over the years: in this early photograph the factory's tenants are Cobb & Ward, building merchants The Pixmore (originally the Nip-In) Café opposite has been serving hungry workers for well over a century.

Works Road

Employees leaving factories on foot and on bicycles, photographed outside the W. H. Smith bookbinding works, opposite the Nip-In Café, at the junction of Works Road and Pixmore Avenue. The buildings remain, but today the car prevails.

St Edmundsbury Weaving Works

Producers of high-quality hand-woven textiles, Edmund Hunter brought his company to this purpose-built Parker and Unwin factory in Letchworth in 1908. St Edmundsbury Weavers produced the town's banner, 'Foursquare Our City' and many other early banners in the town, as well as silks, altar hangings, vestments and tapestries found in churches all over the country.

Icknield Way

Factories on Icknield Way, in a postcard by Julian Taylor in the 1920s.

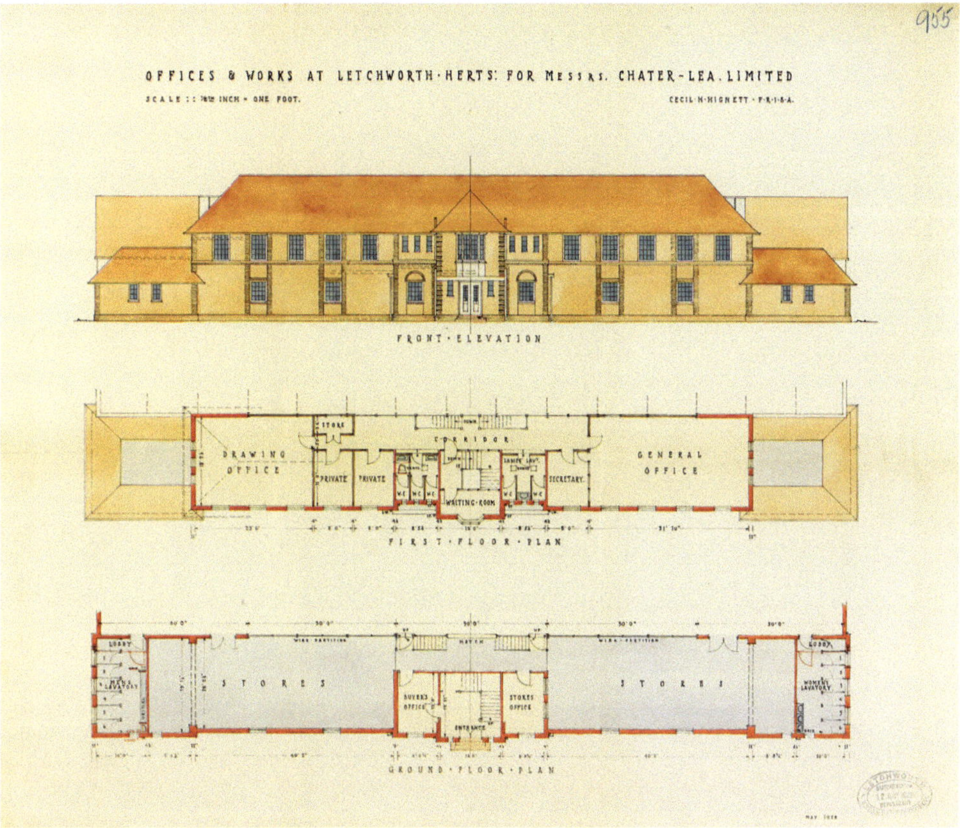

Chater-Lea Factory

Cecil Hignett was one of Letchworth's most under-rated architects. He was prolific, designing buildings of all types from shops and showrooms in the town centre, and thatched cottages and groups of workers housing, and factories like this one for Chater-Lea, who manufactured bicycles and motorcycles here on Icknield Way between 1928 and 1987.

The Spirella Building

Originally from America, the Spirella Company revolutionised the manufacture of corsets by inventing a new type of stay, made from strong, flexible sprung aluminium, rather than brittle whalebone or rigid steel. Cecil Hignett designed this magnificent factory for their UK operation in Letchworth in three phases between 1912 and 1920. After falling into disuse, the factory was sensitively and smartly regenerated in the late 1990s at a cost of some £11 million and today houses modern offices.

The Spirella Gardens

Set in its own beautifully-maintained grounds (then, as now) the building was known as 'the factory of beauty', although the statue in the fountain is sadly no more. The large floor-to-ceiling windows made for well-lit and well-ventilated workspaces.

The Spirella Ballroom

The factory boasted a large assembly hall, or ballroom depending on the function, with chandeliers and a sprung dancefloor, making it perfect for both training conferences (pictured) and social events. The ballroom was brought back to its former glory during the refurbishment of the building and is now hired out for weddings, dances, concerts, conferences and more.

Phoenix Motor Works

A Phoenix car pictured outside the Phoenix Motor works factory in about 1910. Phoenix made vehicles here from 1910 to 1928 when they went out of business, and the factory was taken over briefly by another vehicle manufacturer, Ascot. The factory later became a government training centre later rebranded as a 'Skillcentre', before ultimately being turned into flats in the early 2000s.

Broadway and the Boys' Club and Magistrates' Court

A view along the central promenade of Broadway, leading up to Broadway Gardens, with the Boys' Club (1914) and Magistrates' Court (1918) on the right. This part of Broadway was originally to have accommodated a tram service between Letchworth and Hitchin. Today, the frontages of the Boys' Club and Magistrates' Court have been skilfully incorporated into the Morrisons supermarket which now occupies the premises.

Broadway, the Estate Office and Post Office

Across the road from the Boys' Club and Magistrates' Court, and looking in the opposite direction, is the original estate office for First Garden City Ltd and the post office. The view has changed little even though the use of each building has changed.

Station Place

The Colonnade dominates both the early view and the scene today at Station Place. In 2011 there was a multi-million-pound project to dramatically improve the streetscape in the town centre, including changing the flow of vehicle traffic, new street furniture and planting schemes, repaving throughout in York stone and the water features and sculptures – based on the torches on the town's banner, 'Foursquare Our City' – which can be seen in the modern photo below.

The People's Drapers

This distinctive building played host initially to a number of draper's shops. Originally it was opened in 1906 by George Cramp, and called The People's Drapers, but sadly closed after only a few years. It was taken over by Wightmans in 1909 who operated it successfully until it was taken over by Nicholls (pictured) in the 1930s. In 1960 it became a furniture shop, and then in 1978 a stationers, but has since 2004 been a popular café.

Leys Avenue

This parade of shops on Leys Avenue was designed by local architects, Bennett & Bidwell, and built in 1908. Halfway along the parade was The Simple Life Hotel, a vegetarian temperance hotel and restaurant, catering for Letchworth's early population of idealists, referred to by George Orwell as 'sandal-wearing fruit juice-drinkers' and affectionately known as 'cranks'.

Nott's Bakers

A fixture on Letchworth's other main shopping street, Eastcheap, was Nott's bakers and confectioners, who developed an enviable reputation throughout the district. Today the building is still a bakery.

The Arcade

Designed by Bennett & Bidwell in 1923, The Arcade was as close as Letchworth got to the grand 'Crystal Palace', which Ebenezer Howard outlined in the indicative diagrams for Garden Cities. Nonetheless it was the pride of Letchworth shopping in the 1920s and was beautifully restored in 2010. The Arcade comprises an entire triangular block, from each entrance all the way to the bank on Station Place.

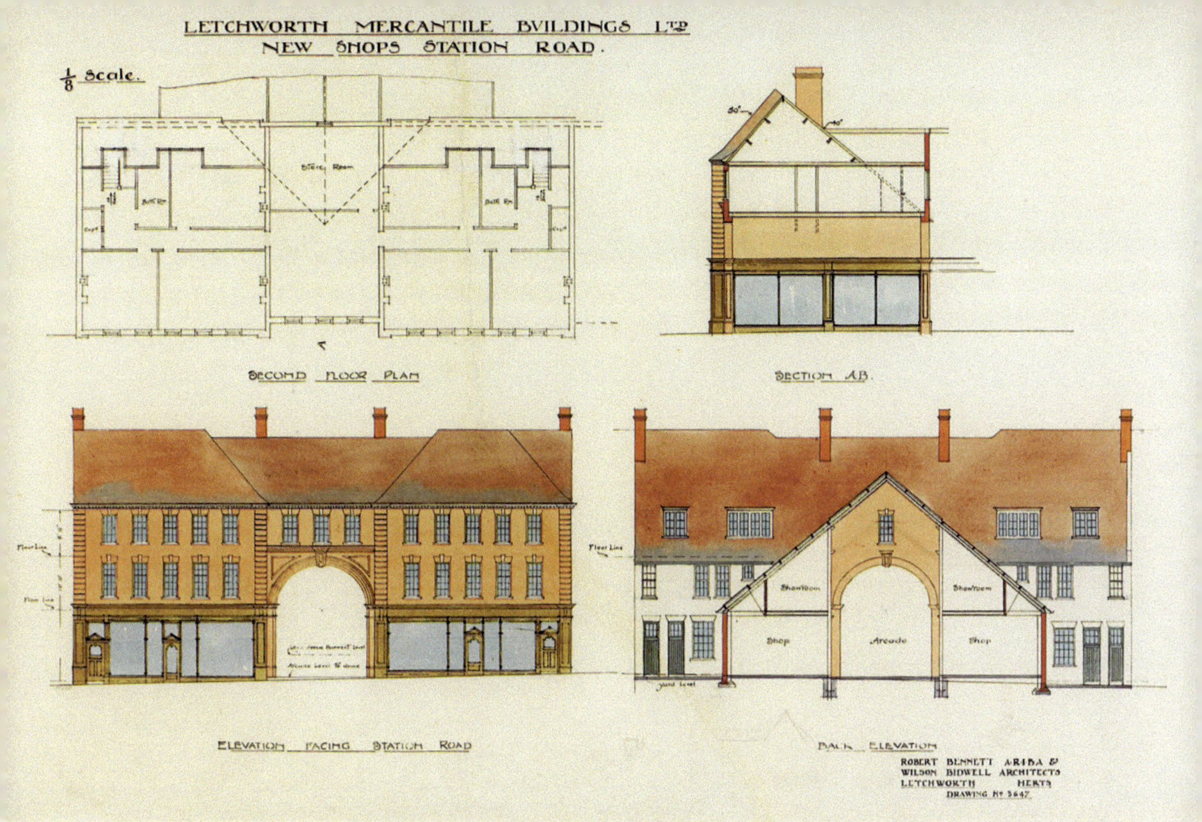

The Arcade Entrance on Station Road

Bennett & Bidwell's plan for The Arcade, rather charmingly called 'Letchworth Mercantile Buildings' reveals that the frontage has changed little from how it was designed.

The Arcade
The Arcade pictured in the 1930s and today.

First Shops, Station Road

The earliest shops in Letchworth developed along Station Road, further down from The Arcade. The first shops built in the town were these, with their small square towers at either end and parapet walls projecting above the roofline between each unit, designed in the London fashion.

Silver Birch Cottages

Across the road from the parade of shops are Silver Birch Cottages, designed by Parker & Unwin in 1905 in typical early Garden City style, with their gables and dormer windows. Today they are very much more successfully screened from the hubbub of Station Road by a century's worth of tree growth, than by the saplings in the original photograph.

Arena Parade

Built in the 1950s, Arena Parade initially provided a very sleek modern addition to the town centre, with red canopies added in the 1980s and more recently there has been a refresh to the corner with the Prezzo restaurant.

The Broadway Cinema

The jewel in the crown of the town centre is Letchworth's magnificent art deco cinema, The Broadway. Designed by Bennett & Bidwell and built in 1936, it was originally a single-screen auditorium with seating for 1,400 people. Today it has been sensitively adapted into a four-screen facility, with the sweep of the balcony being retained in a large (430-seat) Screen 1. Digital projection means that today its offer isn't limited to films but also includes live streaming of opera, theatre, ballet and more.

The Broadway Cinema
Inside the foyer retains its original Italian terrazzo floor and unusual shape. When it was first built the cinema had glittering lights beneath every other seat, an air purification system and smart usherettes in uniforms that matched the lavish peacock-and-gold colour scheme. Owned and operated by Letchworth Garden City Heritage Foundation, at time of publication the cinema is set to enjoy a renovation in 2016 to enable it to put on live entertainment on stage.

Howard Park and Gardens

A long wedge of green space on the original masterplan, Howard Park & Gardens stretches for half a mile or so between Rushby Mead and Norton Way South. Originally a pond was dug by hand by early residents in 1911, before its transformation into a more formal municipal park-style paddling pool in the 1920s. The Gardens had a major re-landscaping in 2011 which added new planting and pathways as well as popular and well-used play areas and water features.

The Grammar School

Designed in 1930 by Barry Parker, the Grammar School served as one of the town's main secondary schools until 1976 when the school was relocated to a modern purpose-built facility on the edge of town, Fearnhill. The Grammar School then served as County Council offices for more than thirty years before being given a new lease of educational life in 2013 when it re-opened as The DaVinci Studio School of Creative Enterprise, for fourteen to nineteen-year-olds.

The Town Hall

The Town Hall was designed by Bennett & Bidwell and opened in 1935. It was something of a compromise on the rather grander plans that had always been intended, but is nonetheless a suitably imposing presence, enjoying an enviable location at the head of Broadway Gardens. North Hertfordshire District Council moved their main offices to a new purpose-built facility in the 1970s, and the Town Hall today is the administrative centre for North Hertfordshire College.

Broadway Gardens

Broadway Gardens was planned to be Letchworth's central civic space. Raymond Unwin's grand plan for the Central Square in 1912 was 'freely adapted from the works of Wren and other Masters' but unfortunately never happened. Instead development consisted of more municipal-style planting schemes from the 1920s onwards with rose gardens and poplars. By the end of the century, the gardens had grown tired and were re-landscaped to celebrate Letchworth Garden City's centenary in 2003, with Unwin's 1912 plan commemorated in a plaque.

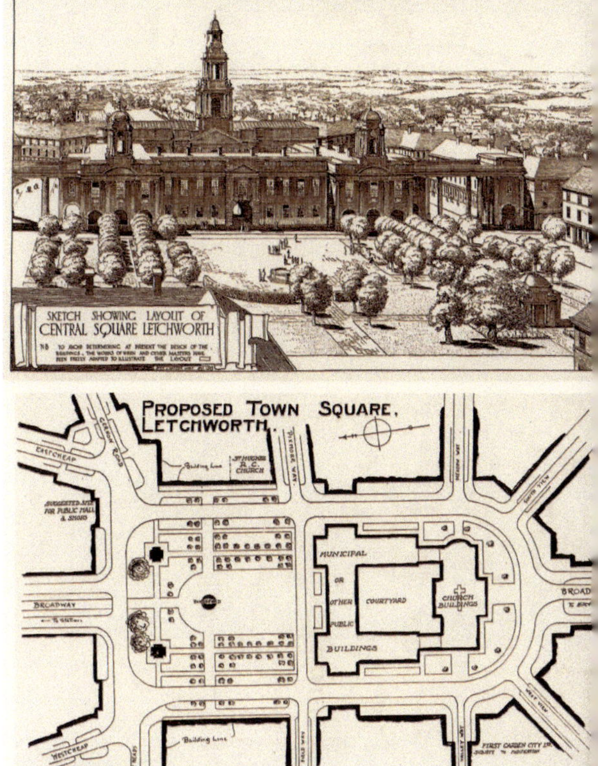

Howard Hall

Letchworth's first public building, The Mrs Howard Memorial Hall was paid for by public subscription and built in 1906, named in honour of Ebenezer Howard's first wife, Elizabeth, who had been a great support to her husband for many years as he struggled to achieve his dream, and who sadly barely saw any of it come to fruition as she passed away in 1904.

Howard Hall

It was designed by Parker & Unwin, loosely based on a building by the Northern Arts & Crafts architect, Edgar Wood, called The Church of the Christ Scientist in Manchester. It was home to, among other things, Letchworth's first parish council, its first lending library and its first dentist. It continues to be used as a social hub to this day and, complete with a sensitive modern extension, remains one of the town's finest early buildings.

The Free Church

Originally the first new church built in Letchworth, paid for and literally built by its early citizens in 1905, the current building was designed by Barry Parker in 1924, (unusually for him) in a neo-Georgian style. Perhaps he was influenced by Louis de Soissons' work at the second Garden City at Welwyn which had started a few years previously.

Howgills

The Society of Friends Meeting House was built in South View in 1907. It was commissioned by Juliet Reckitt, philanthropic daughter of a wealthy Hull industrialist, and designed by Bennett & Bidwell, but loosely based on a seventeenth-century meeting house called Brigflatts, in Yorkshire (and named after the Howgill Hills which surround that hall) but also expertly integrates Arts & Crafts features that are so prevalent elsewhere in Letchworth Garden City.

Howgills

This, one of two identical decorative wooden arches, is a charming feature of the building. Letchworth attracted a wide variety of people with many different beliefs and ideologies, including Quakers, and so a community of Friends was quickly established. Indeed many of the founders of the Garden City movement were Quakers, such as Joseph Rowntree and the Cadbury family, as well as Letchworth architects Benjamin Wilson Bidwell and Thomas Geoffry Lucas.

The Cloisters

One of Letchworth Garden City's more unusual buildings is The Cloisters, a fairy-tale castle-style building commissioned by the eccentric and wealthy Annie Lawrence, and designed by the architect, W. H. Cowlishaw, as an outdoor school. Residential summer schools were held in the early years, starting in 1908. The students slept in hammocks in the open air of the central 'Living Court', swam with frogs in the spring-fed pool and ate at a long marble table.

The Cloisters

The army requisitioned it during the Second World War and returned it in a parlous state. Miss Lawrence tried to find a suitable recipient for the building in her declining years but without very much success. She offered the building to the Freemasons who gratefully accepted the gift. Ownership was transferred in March 1948, and the first Masonic meeting was held on the premises in October 1951. The Cloisters today is still the Masonic Centre for the North Hertfordshire Freemasons.

The Cloisters Fountain

One of the more unusual features of the interior of the building to have survived is this lavabo, or fountain, made of Spanish marble – a centrepiece of the reception hall. The Cloisters organised many events which enhanced the lives of the residents of Letchworth Garden City. Musical and dramatic performances were regular occurrences which drew large crowds, and many groups and organisations used the premises for their meetings.

St Christopher School

Founded in 1915, St Christopher School had its origins in the Theosophy movement, with Annie Besant laying the foundation stone of its original premises (between Spring Road and Broadway, now St Francis College). It was well-known for its progressive teaching methods, not least in offering open-air education. The school moved to these buildings set in acres of land off Barrington Road, originally the (private) Letchworth School, and remains there to this day.

Letchworth Station

Pictured here during its opening in 1913, the town's railway station replaced a temporary wooden structure, which had opened in 1905 to accommodate the throng of visitors to the Cheap Cottages Exhibition that year. One of the advantages of laying out a town from scratch was the ability to locate the railway station right in the heart of the town centre.

Letchworth Station

A view along the platform of the 'new' station in 1913, which has changed little over the years. Letchworth Garden City's location was one of its key advantages, just thirty minutes from both London and Cambridge.

Houses on Ridge Road

Part of the Pixmore Hill estate of early workers housing, these houses on Ridge Road were designed by Parker & Unwin in one of their favourite layouts, the village green design, in groups set back from the road around a communal green space.

Sollershott Circus

Letchworth Garden City is home to many 'firsts' including this, the UK's first roundabout called Sollershott Circus, where six roads converged. It was designed by Parker and Unwin in 1908, and inspired by the Place d'Etoile in Paris, which Unwin had written about a year previously. It was built around 1909, which makes it the earliest in the country.

Norton Road School

The town's first purpose-built school was designed by Raymond Unwin and opened in 1909 (not 1908 as the plaque erroneously says). Later known simply as Norton School, it one of the first 'County' schools in the country. The first elementary school, which it replaced, had been based in 'the sheds' the primitive huts initially built to accommodate the unemployed workers who came to Letchworth to build its roads and services. The school was converted into flats in 2008.

Norton Post Office

The general stores and post office on Norton Road, just to the south of the village of Norton, remain largely unchanged since they were built in *c.* 1908.

Norton Common

Conceived as a central accessible green space – the lungs of the city, like Central Park in New York or the Royal Parks in London – Norton Common today remains a local nature reserve. It is both a popular destination for recreation and leisure, and a verdant walking route from the north of Letchworth towards the town centre. This avenue of trees follows the same axis on the map, from south-west to north-east, as Broadway.

Letchworth Corner Post Office

Situated on the main thoroughfare between Letchworth's neighbouring towns, Hitchin and Baldock, this pretty building was once the town's sub-post office. Today it is a shop serving locals and passers-by and a steady stream of pupils from the nearby St Christopher School.

Broadway

Letchworth Garden City's formal boulevard, Broadway. This section, between the Town Square and the first roundabout is double tree-lined, belying its purpose as a grand entrance to the town. The view is rather different today from the quiet original over a hundred years ago.

Hitchin Road

The main road between Hitchin and Baldock as it passes through Letchworth, here near South View and Sollershott West (to the right).

The Liberal Catholic Church of St Alban

This little church, on the corner of Norton Way South and Meadow Way, was built in 1923 and consecrated in 1924. Early Letchworth Garden City was a tolerant melting pot of different religions, and there a large number of different surviving churches as a result, although not all are as charming as this one.

Rushby Walk Roofs

Nothing typifies the classic early 'Letchworth Look' design of much of the town's early houses as the rooflines here at Rushby Walk.

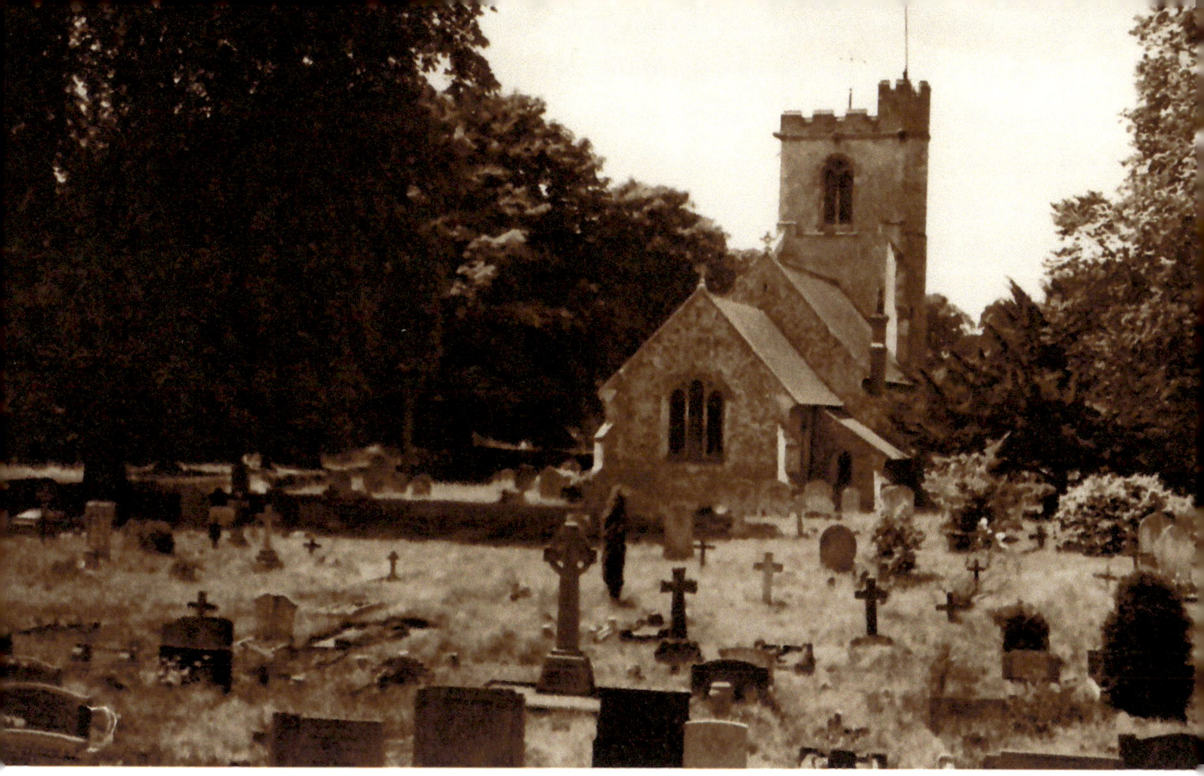

All Saints Church, Willian

Parts of the original parish church at Willian – All Saints – date back as much as 900 years. Its tower has clocks on three sides, the fourth side was left blank, supposedly so that farm labourers would not be able to tell when to leave work.

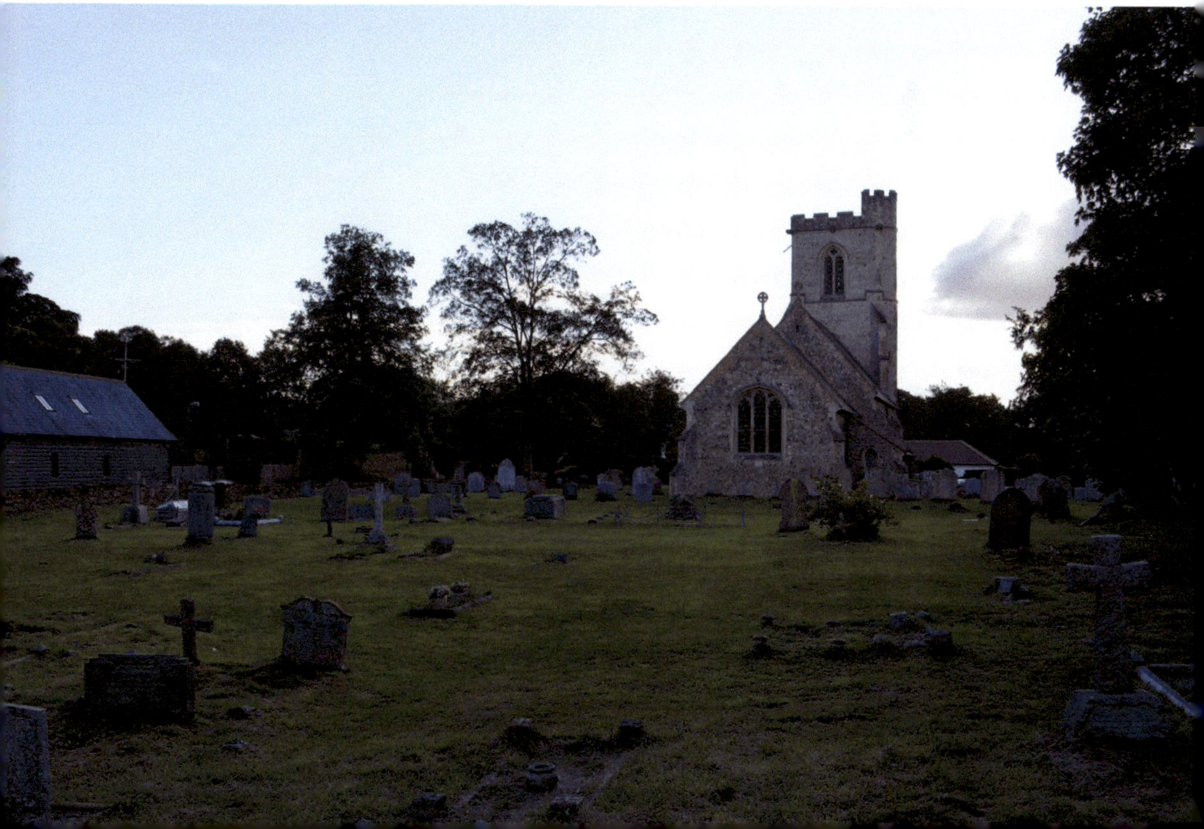

Letchworth Garden City From Above

The first aerial view shows the town in about the late 1940s. The more recent photograph, taken in August 2013, from a different orientation, shows clearly how verdant Letchworth Garden City is today, and how closely the original masterplan was followed. It seems an appropriate point at which to end this journey around Letchworth Garden City through time.

The social experiment has thrived and survived, and is in good shape to look forward to another hundred years and more.

Acknowledgements

All old photographs © The Garden City Collection, Letchworth Garden City Heritage Foundation www.gardencitycollection.com

All modern-day photographs © Josh Tidy, except bottom image page 5, bottom image page 8, bottom image page 73 and bottom image page 94, © Letchworth Garden City Heritage Foundation.

About the Author

Josh Tidy is the Curator at the The International Garden Cities Exhibition, a museum and visitor centre funded and run by Letchworth Garden City Heritage Foundation. He has a keen knowledge of the history of Letchworth and the Garden City Movement and is a passionate advocate for Garden Cities past and present across the world. He has been celebrating and sharing this special town's history since its centenary in 2003.